PHYSICS MATTERS

PHYSICS MATTERS

Vasant Natarajan
Indian Institute of Science, India

World Scientific

NEW JERSEY · LONDON · SINGAPORE · BEIJING · SHANGHAI · HONG KONG · TAIPEI · CHENNAI · TOKYO

Published by

World Scientific Publishing Co. Pte. Ltd.

5 Toh Tuck Link, Singapore 596224

USA office: 27 Warren Street, Suite 401-402, Hackensack, NJ 07601

UK office: 57 Shelton Street, Covent Garden, London WC2H 9HE

Library of Congress Cataloging-in-Publication Data
Names: Natarajan, Vasant, 1965– author.
Title: Physics matters / Vasant Natarajan (Indian Institute of Science, India).
Description: Singapore ; Hackensack, NJ : World Scientific Publishing Co. Pte. Ltd., [2016] |
   Includes index.
Identifiers: LCCN 2016023759 | ISBN 9789813142503 (hc ; alk. paper) |
   ISBN 9813142502 (hc ; alk. paper) | ISBN 9789813142510 (pbk ; alk. paper) |
   ISBN 9813142510 (pbk ; alk. paper)
Subjects: LCSH: Physics.
Classification: LCC QC21.3 .N38 2016 | DDC 530--dc23
LC record available at https://lccn.loc.gov/2016023759

British Library Cataloguing-in-Publication Data
A catalogue record for this book is available from the British Library.

Desk Editors: V. Vishnu Mohan/Low Lerh Feng

Typeset by Stallion Press
Email: enquiries@stallionpress.com

Printed in Singapore

To

Amma and Appa

If I have seen further than others, it is by standing upon the shoulders of giants.

— Isaac Newton

Contents

Abbreviations

AC	—	Alternating Current
AOM	—	Acousto-Optic Modulator
BEC	—	Bose–Einstein Condensation
CCD	—	Charge-Coupled Device
DC	—	Direct Current
EOM	—	Electro-Optic Modulator
FM	—	Frequency Modulation
FWHM	—	Full-Width-at-Half-Maximum
GPS	—	Global Positioning System
MOT	—	Magneto-Optic Trap
NMR	—	Nuclear Magnetic Resonance
QED	—	Quantum Electro-Dynamics
RF	—	Radio Frequency
SNR	—	Signal-to-Noise Ratio
SOF	—	Separated Oscillatory Fields
TOP	—	Time-Orbiting Potential

Preface

SCIENCE is universal—it has no boundaries. This is especially true in today's internet connected world. Scientific results travel across the globe at (literally) lightning speed, because of the optical fibers connecting computers worldwide. This not only makes the latest scientific advances accessible to researchers in any corner of the world, but also means—as the number of scientists increases exponentially—that there is a lot of material that one must imbibe before any real progress can be made. In effect, the foundational basis of science in general (and physics in particular) has become bigger and bigger. Therefore, standing on the shoulders of past giants and seeing further requires more and more effort. Gone are the days of an Einstein working alone in a patent office and making fundamental contributions to physics. Einstein could do this because most of the important results in physics known before his time could be imbibed through a few months worth of study.

Modern science, and the scientific method of attributing cause-effect relation to any result, is a relatively recent development. It dates back to the start of the *Renaissance* period in Europe a few centuries ago, and certainly no earlier than the time of Galileo's famous experiments at the Leaning Tower of Pisa in 1589. The resulting science of mechanics—**classical mechanics** as it is known today—is embodied in Newton's three laws of motion. The nice thing about these laws (and the observational results of Galileo, Kepler, Copernicus, and others, on which they were based) is that they were presented in a common language accessible to other scientists—Newton's *Philosopiae Naturalis Principia Mathematica* is in Latin, the common language of science at that time.

Of late, the study of science has become highly compartmentalized, with the result that students of other subjects—both science and arts—do not study physics in the required detail. But I strongly feel that there are many concepts in physics that have wide applicability, and therefore should be understood by all students. Understanding concepts in other subjects also encourages out-of-the-box thinking, and making original contributions to one's chosen field. This book aims to fulfill this need by presenting important concepts in physics in a manner that is accessible to non-physics students. The use of equations is necessary in physics—*mathematics is the language of science*—but they are kept to a minimum. The last few pieces are on recent Nobel Prizes in Physics given for areas related to Atomic and Optical Physics—my area of research—which shows the relevance of this area of physics to modern society.

Acknowledgments

I thank my assistant S Raghuveer for help with the typing of the manuscript; student Sumanta Khan for help with the figures; and student Apurba Paul for help with typesetting in LaTeX. Chapters 2–5 and 7–12 are expanded from earlier articles that appeared in *Resonance – Journal of Science Education* (© Indian Academy of Sciences), and are reproduced here with permission.

Vasant Natarajan
Bangalore, 2016

About the Author

Vasant Natarajan did his B.Tech. from Indian Institute Technology, Madras; his M.S. from Rensselaer Polytechnic Institute, Troy, New York; and his Ph.D. from MIT, Cambridge, Massachusetts. He then worked for two years at AT&T Bell Labs, Murray Hill, New Jersey. He joined the Physics Department at IISc in 1996, where he has been ever since. His research interests are in laser cooling and trapping of atoms; quantum optics; optical tweezers; quantum computation in ion traps; and tests of time-reversal symmetry violation in the fundamental laws of physics.

Chapter 1

Oscillations

A LMOST all undergraduate textbooks on physics treat the simple harmonic oscillator in detail because it is the key to understanding much of physics. Apart from direct applications such as in a pendulum clock, it can be used to understand collective excitations like phonons in a solid. But perhaps the most important reason to study them is that it is first step in understanding light. A light wave is like a harmonic oscillator, except that instead of position and velocity oscillating as in a mechanical oscillator, it is the electric and magnetic fields that oscillate in a light wave. In addition, as we will see in this chapter, it can be used to understand many phenomena that are normally associated with quantum mechanics. The reason for this is that the main equation governing quantum mechanics is the Schrödinger equation, which is just a modified wave/oscillator equation.

A. The simple pendulum

We first analyze the dynamics of the simple pendulum, mainly because it is slightly different from the normal application of Newton's second law of motion $F = ma$. As shown in Fig. 1.1, the dynamical variable in the pendulum is the angle θ and not the position x. This implies that the acceleration has dimensions of $[T]^{-2}$, which means that the dimensions of the inertial term has to be $[M][L]^2$ in order to get the dimensions of force correct. The modified equation of motion is therefore

$$\tau = I \frac{d^2\theta}{dt^2}$$

where τ is the torque (moment of force) about the point of contact, and $I = m\ell^2$ is the moment of inertia of the pendulum bob about the point of contact. The torque is given by

$$\tau = \vec{r} \times \vec{F} = -mg\ell \sin\theta$$

The minus sign is because the torque is in the $-\theta$ direction.

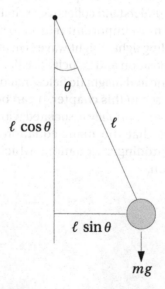

θ

$\ell \cos\theta$

ℓ

$\ell \sin\theta$

mg

Figure 1.1: Simple pendulum showing relevant parameters used in the analysis.

Using the small angle approximation that $\sin\theta \approx \theta$ in the equation of motion, we get

$$m\ell^2 \frac{d^2\theta}{dt^2} = -mg\ell\theta$$

$$\implies \quad \frac{d^2\theta}{dt^2} = -\frac{g}{\ell}\theta$$

This is the familiar harmonic oscillator equation of motion. Its general solution—which can be verified by substitution—is given by

$$\theta(t) = \theta_\circ \sin(\omega t + \varphi) \tag{1.1}$$

Here $\theta_\circ$ is the amplitude of the motion, φ is the phase at $t = 0$, and

$$\omega = \sqrt{\frac{g}{\ell}} \tag{1.2}$$

is the (angular) frequency of the oscillator. $\theta_\circ$ and φ are determined by the initial conditions, i.e. initial position and initial velocity of the bob—**two** constants in the solution because the equation of motion is a **second** order equation.

The defining characteristic of harmonic motion is that the frequency is independent of amplitude. In the case of the simple pendulum, this is true only when the angle is small, so that $\sin\theta$ can be approximated as θ. The time period of motion T (defined as the time taken for one complete cycle) is written as

$$T = \frac{2\pi}{\omega} \approx 2\pi\sqrt{\frac{\ell}{g}}$$

independent of amplitude $\theta_\circ$. For larger amplitudes, T becomes a function of $\theta_\circ$, and the oscillator is called "anharmonic". A Taylor expansion of the sin term in the equation of motion then yields

$$T = 2\pi\sqrt{\frac{\ell}{g}}\left(1 + \frac{1}{16}\theta_\circ^2 + \cdots\right) \tag{1.3}$$

which shows explicitly that the time period increases at larger amplitudes.

This behavior can be understood from the shape of the potential energy curve. For a conservative force like that due to gravity, the force can be derived from a potential U as follows

$$F = -\frac{dU}{dx}$$

Geometric analysis of the force on the pendulum bob shown in Fig. 1.1 gives for the potential

$$U = mg\ell(1 - \cos\theta)$$

This potential (with $mg\ell = 1$) is shown in Fig. 1.2 for a range of $-\pi$ to $+\pi$. A Taylor expansion of the potential around $\theta = 0$ yields

$$U = mg\ell\left(\frac{\theta^2}{2!} - \frac{\theta^4}{4!} + \cdots\right)$$

The first term in this expansion is the parabolic or harmonic term, for which the small angle approximation is valid. For comparison, this parabolic shape is also shown in the figure (as a dotted line). It is seen that the real potential is shallower than the harmonic one.

The motion in the full potential will remain periodic, i.e. return to the same point in the potential curve, because the system has no way of losing energy. However, it will not be sinusoidal—have a single frequency component—as in the case of a harmonic oscillator. Furthermore, in accordance with Eq. (1.3), the time period will increase with amplitude.

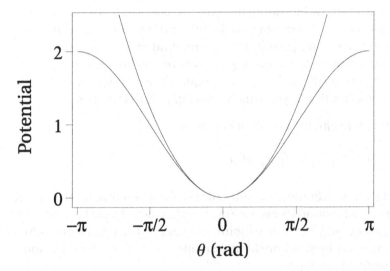

Figure 1.2: Shape of the potential seen by the pendulum bob. For comparison, the parabolic approximation near $\theta = 0$ is shown with a dotted line.

B. Parametric amplification

Parametric amplification is a method to amplify the motion by modulating some parameter of the system at twice the natural frequency. This is different from normal amplification, where the amplitude is increased by a periodic driving force at the oscillation frequency.

The idea behind parametric amplification can be understood by considering the motion of a child on a swing. After playing on it for a while, she learns to increase her amplitude by instinctively crouching at the bottom—thereby increasing the length of the swing—and stretching at the ends—thereby decreasing the length of the swing—thus changing the length twice every period. This is a method of self-amplification which does not require contact with ground or boosts from another person. The parameter of the swing (or the equivalent pendulum) that is being changed at is the length, which from Eq. (1.2) determines ω.

Mathematically, the potential becomes

$$U = \frac{1}{2}mg\ell\theta^2 + \frac{\varepsilon}{2}mg\ell\theta^2 \sin 2\omega t$$

the first term is the normal harmonic oscillator potential that we saw earlier. The second term arises because the length is modulated at twice the natural frequency—i.e. frequency modulation at 2ω. The degree of modulation, determined by ε, is considered small—i.e. $\varepsilon \ll 1$. The corresponding equation of motion is

$$\frac{d^2\theta}{dt^2} + \omega^2\theta = -\omega^2\theta\varepsilon \sin(2\omega t) \tag{1.4}$$

since ε is small, we get the solution

$$\theta(t) = B(t)\cos\omega t + C(t)\sin\omega t$$

this is the standard solution for the motion of a pendulum, as can be verified by expanding the solution in Eq. (1.1), except that B and C are not constants but functions of time. But their time variation is taken to be slow enough that the second derivative can be neglected. Therefore

$$\frac{d^2\theta}{dt^2} \approx -\omega^2\theta(t) - \omega\frac{dB}{dt}\sin\omega t + \omega\frac{dC}{dt}\cos\omega t$$

Substituting into the equation of motion in Eq. (1.4), we get

$$-\omega \frac{dB}{dt} \sin \omega t + \omega \frac{dC}{dt} \cos \omega t = -\omega^2 \varepsilon \left[B \cos \omega t + C \sin \omega t\right] \sin (2\omega t)$$

Using trigonometric identities and averaging away the terms rapidly oscillating at 3ω, yields

$$-\frac{dB}{dt} \sin \omega t + \frac{dC}{dt} \cos \omega t = -\frac{\varepsilon \omega}{2} \left[B \sin \omega t + C \cos \omega t\right]$$

The coefficients of the sin and cos terms must be separately equal, which gives

$$\frac{dB}{dt} = +\frac{\varepsilon \omega}{2} B \qquad \text{and} \qquad \frac{dC}{dt} = -\frac{\varepsilon \omega}{2} C$$

Solving the above yields

$$B(t) = B_\circ e^{+\varepsilon \omega t/2} \qquad \text{and} \qquad C(t) = C_\circ e^{-\varepsilon \omega t/2}$$

This shows that the amplitude of the cos quadrature increases exponentially with time, while the sin quadrature decreases by the same factor. The amplitude of the motion

$$\theta_\circ(t) = \sqrt{B^2(t) + C^2(t)}$$

increases with time—the signature of any amplification process.

1. Squeezed states

The above discussion of parametric amplification leads us to another concept in quantum mechanical states of light, namely squeezed states. The light wave, like a harmonic oscillator, has two degrees of freedom—the sin and cos components of the electric (or magnetic) field. Each of these has a mean squared value determined by the Heisenberg uncertainty principle, which are equal for the ground or vacuum state. The uncertainty distribution in phase space—with sin and cos axes—is circularly symmetric. This can be "squeezed" along some axis, so that the distribution becomes elliptic while maintaining its area. Such squeezed states are important for precision measurements because they can be used to go below quantum noise limit set by the uncertainty principle.

The idea of squeezing is entirely analogous to the statistical distribution of an ensemble of identical harmonic oscillators in thermal equilibrium with a reservoir. The thermal reservoir gives kicks to the oscillators, so that the statistical distribution in each quadrature is that required by thermal equilibrium. From the above analysis, we see that the sin quadrature represents the motion of the pendulum, while the cos quadrature represents its velocity. Each of these varies from 0 to ∞ with an average value of $k_B T/2$, as determined by the equipartition theorem. Thus the distribution in phase space is circularly symmetric. On applying the parametric drive, the distribution becomes elliptic, because the cos quadrature increases while the sin quadrature decreases. Since parametric excitation is a coherent non-dissipative process, in accordance with Liouville's theorem the area in phase space is conserved.

Hence, generation of a squeezed statistical distribution of classical oscillators is entirely analogous to the generation of squeezed states of photons.

C. Ramsey separated oscillatory fields method

The SOF technique is one of the most powerful methods of precision spectroscopy. Though it is now a technique used in all kinds of high-resolution frequency measurements, it was originally conceived by Norman Ramsey as a technique for improvement to the RF resonance measurements pioneered by Rabi. As the name suggests, it involves the sequential application of transition producing oscillatory fields to the system under consideration, with an interval in between. This was experimentally realized by Ramsey by having two separated resonance coils through which a molecular beam passed sequentially.

In effect, the method represented the first exploitation of interference between two parts of a quantum superposition state. This will become clear by seeing how Ramsey came up with this idea. Ramsey, who was Rabi's Ph.D. student and therefore thinking about improving Rabi's resonance method in the background, was explaining to UG students at Harvard University that the resolution of an optical telescope can be improved by blackening the central portion of the primary lens. This is because the lens forms an image by taking a spatial Fourier transform of the object—blackening the central portion results in removing the low-frequency components from the center, while keeping the high-frequency components near the edge which contributes to the resolution by interference. In the SOF method, the blackening of the central portion corresponds to having a dark region between the two OF regions. Of course, this darkening will result in a loss of overall signal strength—this kind of trade-off between signal strength and resolution is a common feature of precision measurements.

The SOF method is now regularly used in the most precise atomic clocks around the world. The basic idea can be understood by considering a child moving back and forth on a swing and measuring her oscillation frequency. Measuring the frequency is similar to keeping the child swinging by giving periodic pushes. The dark period in the SOF technique corresponds to letting the child swing freely between two successive pushes. During the time between the two pushes, the frequency difference between the two oscillators—the child swinging and the person pushing—builds up as a phase difference, so that after a sufficiently long dark time the person pushing is exactly out of phase with the child, and the second push brings the child to a complete halt. Thus, even a small frequency difference can be built up to a large phase difference simply by measuring the dark period—the frequency mismatch can be measured with increasing precision by waiting for ever longer times between the two pushes. This is the advantage in the latest Cesium atomic clock—by using laser cooled atoms the dark period is about 1 s, whereas it was about 1000 times smaller in older atomic clock with thermal beams.

D. Coupled oscillators

Understanding coupled oscillators is essential to understanding several concepts in physics. The simplest coupled oscillator system consists of two identical—same frequency—pendula suspended from a common support. As discussed earlier in this chapter, the pendula can be made identical by having the same length—independent of the shape, size, or mass of the bob. The common support provides a coupling, which changes the equations of motion for the two displacements θ_1 and θ_2 as follows

$$\frac{d^2\theta_1}{dt^2} = -\frac{g}{\ell}\theta_1 - \epsilon(\theta_1 - \theta_2)$$

$$\frac{d^2\theta_2}{dt^2} = -\frac{g}{\ell}\theta_2 - \epsilon(\theta_2 - \theta_1)$$

The second term in each of the above equations is due to the coupling, whose strength is determined by the parameter ϵ. The presence of this term makes the equation unsolvable for θ_1 and θ_2 independently. However, there are linear superpositions of these displacements—called *normal modes*—which obey the harmonic oscillator equations of motion, and can be solved independently to find the frequency of each normal mode.

Instead of finding the normal modes in a general manner applicable to all coupled oscillator systems, we guess the following

$$\eta_c = (\theta_1 + \theta_2)/\sqrt{2} \qquad \text{Common mode}$$

$$\eta_s = (\theta_1 - \theta_2)/\sqrt{2} \qquad \text{Stretch mode}$$

Substituting back into the equations of motion, we get

$$\frac{d^2\eta_c}{dt^2} = -\frac{g}{\ell}\eta_c$$

$$\frac{d^2\eta_s}{dt^2} = -\left(\frac{g}{\ell} + 2\epsilon\right)\eta_s$$

This shows explicitly that the normal modes are independent, and have oscillation frequencies of

$$\omega_c = \sqrt{\frac{g}{\ell}} \qquad \text{for the common mode}$$

$$\omega_s = \sqrt{\frac{g}{\ell} + 2\epsilon} \qquad \text{for the stretch mode}$$

What this means is that if either one of the normal modes is excited, then the motion stays in that mode forever. On the other hand, if only the first oscillator is excited (by giving θ_1 an initial amplitude), then both normal modes are excited because

$$\theta_1(0) = 1 \quad \text{and} \quad \theta_2(0) = 0$$

$$\implies \quad \eta_c(0) = \frac{1}{\sqrt{2}} \quad \text{and} \quad \eta_s(0) = \frac{1}{\sqrt{2}}$$

The time dependence of each normal mode is the standard harmonic oscillator solution given by

$$\eta_c(t) = \eta_c(0)\cos\omega_c t \quad \text{and} \quad \eta_s(t) = \eta_s(0)\cos\omega_s t$$

We can now project back to the individual displacements as follows

$$\theta_1(t) = \frac{1}{2}(\cos\omega_c t + \cos\omega_s t) = \cos\left(\frac{\omega_c + \omega_s}{2}t\right)\cos\left(\frac{\omega_s - \omega_c}{2}t\right)$$

$$\theta_2(t) = \frac{1}{2}(\cos\omega_c t - \cos\omega_s t) = \sin\left(\frac{\omega_c + \omega_s}{2}t\right)\sin\left(\frac{\omega_s - \omega_c}{2}t\right)$$

These two motions are shown in Fig. 1.3. The rapid oscillation seen for θ_1 and θ_2 corresponds to the oscillatory motion of each displacement at the natural frequency. It should be obvious that the amplitude will not change with time in the absence of coupling. However, in the presence of coupling the amplitude shows a slow variation—an amplitude modulation at the frequency difference between the two normal modes. As expected, the amplitude of the first oscillator goes from a maximum to zero to a maximum again, and then the pattern repeats indifferently. The other bob shows the same behavior but out of phase with the first, namely zero to a maximum to zero again, and so on. The zero means that the amplitude swaps completely between the two oscillators, which happens only if the uncoupled oscillators are identical, or degenerate in the language of physics.

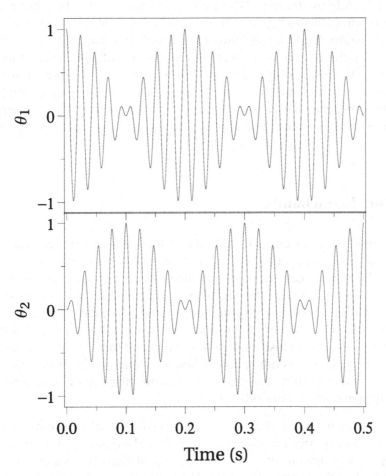

Figure 1.3: Oscillations in the motion of two coupled (identical) pendula. The first one is started with initial amplitude (1 in some units). $\theta_1(t)$ and $\theta_2(t)$ then show out-of-phase amplitude modulation at the Rabi frequency. There is complete energy transfer because the two oscillators are identical.

1. Rabi oscillations

The above analysis leads to a widely used concept in atomic physics, namely *Rabi oscillations*. When two levels are coupled by a laser, the population oscillates between the two levels, exactly akin to the manner in which the amplitude swaps between two coupled oscillators. This so-called "Rabi flopping" occurs at the Rabi frequency—the difference in frequency between the two normal modes, as shown before. If the coupling is tuned on for exactly the length of time needed for a complete swap, then it is called a "pi pulse". This kind of pulse is immensely useful in NMR, where it can be used for a complete population transfer between two magnetic levels.

2. Bonds and bands

The underlying reason for the difference in frequencies between the two normal modes is the difference in symmetries of the two modes in terms of the original variables—the common mode is the symmetric combination, while the stretch mode is the antisymmetric combination. The symmetric combination always has lower frequency (and hence lower energy by the relation $E = h\nu$) compared to the antisymmetric one. In chemical bonding, if two atoms come close enough to interact and form molecular bonds, then the symmetric combination of their wavefunctions leads to the lower-energy bonding orbitals, whereas the antisymmetric combination leads to the higher-energy antibonding orbitals.

In a crystalline solid, the orbitals extend to a quasi-continuous band because of the large number of atoms involved—typically of the order of Avogadro's number of 10^{23}. The bonding orbital then becomes the valence band, while the antibonding orbital becomes the conduction band. The energy gap between the two is called the *bandgap* and is an important parameter in semiconductor devices. As discussed before, the bandgap is a measure of the coupling between the atom. Therefore, it can be changed in either of the following ways

1. Heating the solid, which increases the interatomic spacing because of thermal expansion, and hence reduces the coupling.

2. Applying pressure to the solid, which reduces the interatomic spacing, and hence increases the coupling.

This shows that the bandgap of a solid is not a predetermined quantity, but one which is subject to experimental manipulation.

3. Spontaneous decay

The coupled oscillator idea can also be used to make an elegant model for spontaneous decay. In this model, the oscillator is coupled to a large number N of (non-identical) oscillators, with N tending to ∞. From our previous analysis, we know that the amplitude (or energy) will decay from the main oscillator to the reservoir oscillators, but will return back to the main oscillator after some time. However, this time will become larger and larger as N becomes large.

One might correctly wonder as to where a large reservoir of oscillators is present. We clarify that this is from vacuum fluctuations of the electromagnetic field, which is a necessary prediction of quantum mechanics. When any oscillator is quantized, the lowest energy state is not one of complete rest—as it would violate the Heisenberg uncertainty principle—but one in which the position and momentum have residual fluctuations. Applied to the quantized electromagnetic field, this means that the vacuum (or 0 photon) state has fluctuations of the electric and magnetic fields. In order to maintain consistency with the fact that it is a zero photon state, the average value of the E and B fields is zero, but their mean squared value is not zero.

The model of spontaneous decay is now easy to understand. An atom in the excited state decays to the ground state by coupling to the large number of vacuum modes. However, our previous analysis shows that this is not a one-way process, but one where the atom will eventually come back to the excited state. The time for going back to the excited state is called the *revival time*. Since the number of vacuum modes tends to ∞, the revival time also tends to ∞. Thus, consistent with experimental observations, an atom that has decayed to the ground state will never revive back in real time. This model also predicts that the rate of spontaneous decay can be reduced by decreasing the number of available vacuum modes—a fact which has been demonstrated experimentally using what are called photonic bandgap materials, or control photon fields.

Chapter 2

The myth of cell phone radiation

T HIS chapter is devoted to an analysis of the radiation that emanates from cell phones, and how they cannot cause cancer. Understanding this is important because the number of cell-phone users worldwide has increased exponentially in the past few years. Students have a moral obligation to understand these ideas, and explain it to other users.

A. Photoelectric effect

Albert Einstein won the Nobel Prize in Physics not for his work on Relativity but for his explanation of the *Photoelectric Effect*. This explanation was considered revolutionary enough for a Nobel because it was the first independent confirmation of the paradigm-shifting quantum concept introduced by Planck a few years earlier, which stated that a light wave could only carry energy in *discrete* packets.

The photoelectric effect is the phenomenon where light incident on a metal (or some other surface) causes electrons to be emitted. It had been studied for quite some time before Einstein came along, and there were two observations that were inexplicable from the classical wave picture of light:

1. Photoelectrons were emitted only if the incident light had a frequency above a threshold level, *independent of the intensity*.

2. The *number of photoelectrons* produced above threshold was proportional to the light intensity.

They were inexplicable from the classical picture because, assuming that there was a threshold energy that had to be overcome before electrons were emitted, one could always reach this requirement for a classical wave by suitably cranking up its intensity.

In comes Einstein and the photon picture. With the ideas that the energy per photon is quantized in units of its frequency, and that one needs a single photon with sufficient energy to produce a photoelectron, it is simple to see that there would be a threshold frequency for the effect. In addition, the number of photoelectrons would be proportional to the number of photons in the EM field, i.e. its total intensity. Thus, Einstein could explain all the observations of the photoelectric effect with the *reasonable* assumption that the transition involved in emitting an electron is mediated by one photon of suitable energy. It is reasonable because the transition is from one energy level (where the electron is bound) to another energy level (where the electron is free). There are no other levels in between, which if present could be used as "stepping stones". This explanation is so elegant and simple to understand that it is presented to high-school students in textbooks today. But let us not forget how radical it was when it was first proposed 100 years ago. And how much of a departure from the accepted notions about light.

B. Radiation and cancer

This is now our accepted understanding of all *bond-breaking* processes. Every such process involves a transition with a single photon of sufficient frequency (or energy), and a million photons of *sub-threshold* frequency cannot cause the transition. Or a billion. Think of it like this. If you had a cannon that could shoot a cannonball to a distance of 1 km, 10 cannons will not allow you to hit a target that is 10 km away. Cannon ranges do not add. Similarly, if you could leap a distance of 10 feet, you could jump across a stream that was 10 ft wide. But 9 additional people with the same ability cannot help you cross a 100 ft wide stream.

Which brings to the question of cell-phone radiation and its purported link to cancer. Cancer, *of the kind mediated by radiation*, is known to be caused by mutations in the cell-division machinery—a clear bond-breaking process—which results in uncontrolled multiplication of the cells. X-rays are well known to cause such mutations, which is why X-ray technicians are required to wear lead aprons. UV rays from the sun, those which are not stopped by the ozone layer, can cause skin cancers in people who do not have enough pigmentation to block them. That is why fair-skinned people have to use UV-blocking creams before going out into the sun. But visible light *cannot* cause such mutations. It is *sub-threshold*. And so is any EM wave whose frequency is smaller—such as infrared, microwave, radio waves, and the typical waves (~900 MHz) used for cell phones. This means that the cell-phone photons do not have enough energy to cause a mutation in your DNA. Period. No matter what their power is—increasing their power will increase the number of photons, but they will all be below the threshold for causing cancer. They do not have enough energy to break a bond and cause a mutation.

A skeptic might argue that bond-breaking mutations are not the only way to cause cancer. True. *Heat* can cause damage to living tissues. And definitely if you give enough photons of sub-threshold frequency, you can heat a substance. That is why you feel hot when you go out into the sun. The sub-threshold visible and infrared photons heat up your body, *but they do not cause any damage that can lead to cancer*. And the heating happens because the power density from the sun received on the surface of the earth—called the "insolation"—is typically 1000 W/m^2, while that at the base of a cell-phone tower is ten thousand times smaller or about 0.1 W/m^2.

No wonder you do not feel hot when you stand next to a cell-phone tower.

And this is exactly how a microwave oven works. It heats up the food inside by bombarding it with microwave photons. These photons have a typical frequency of 2.45 GHz, or two and a half times that of cell phones. But even with the higher single-photon energy, the power level inside the oven required for it to heat the food is about 700 W. To understand this scale, consider that the energy inside an oven in one second is equal to that got by using a cell phone *continuously for several days*. Furthermore, since a small fraction of the microwave photons come out of the oven, you actually get a larger exposure by standing next to an oven than from a cell phone. But nobody worries about it because the photons are known to be harmless. Otherwise, microwave ovens would not be so commonly used today. Another fact that is useful is that the cell phone runs off a small battery for several days, whereas the microwave oven is the biggest electricity guzzler in the house using several kilowatts of power. The small cell phone just does not have enough energy to cause significant heating, let alone any tissue damage.

Because we evolved to live in the sunny plains of Africa, our bodies have another defense against non-ionizing radiation, namely a layer of dead cells on the outermost part of our skin. Most radiation does not make it past this layer, which is where it is absorbed to make us feel hot. Therefore, you can be sure than any radiation from the cell phone will not penetrate into the body. In addition, our brains are designed so that they do not overheat, by circulating blood as a coolant. If the bright sun cannot overheat your brain, do you think a small cell phone pressed against your ear can?

C. Epidemiological studies

Despite the overwhelming "physics" evidence that cell-phone radiation is harmless, medical organizations like the WHO have to base their recommendations on "epidemiological studies"—studies that compare the prevalence of cancer or other health indicators between cell phone users and non-users. This is because there are scare-mongers who will play on our fears by giving *anecdotal evidence* that someone who developed brain cancer "was always talking on the cell phone", and therefore the radiation from the cell phone *caused* the cancer. This is a well-known logical fallacy called *post hoc ergo propter hoc,* meaning that if A follows B then A was caused by B. To establish causation, the very least one must show is that no B also implies no A. And this is exactly what epidemiological studies do, they see if there is a causal difference in the prevalence of some health indicator between users and non-users. And the difference should be statistically significant—achieved by studying a large number of people, and not just one or two.

In any case, all of us (cell-phone users) are unwittingly part of the largest epidemiological study ever undertaken in the history of mankind. The total number of cell-phone users in the world is now an unprecedented 80% of the population, up by a factor of 1000 from 20 years ago. *But there is no correspondingly large increase in the prevalence of cancers, especially those which could be caused by cell phones, during that time.* Any ill effects of cell-phone usage would surely have shown up by now in the billions of users worldwide.

Chapter 3

What Einstein meant when he said "God does not play dice ..."

Einstein's famous statement that "God does not play dice with the universe" has two misconceptions in the popular interpretation—his use of the words God and dice—which are corrected in this chapter. After reading this chapter, the reader should hopefully come away with a deeper understanding of Einstein's mind.

A. Einstein—An atheist

The first misconception associated with the statement is that his use of the word 'God' implies that he was a religious person who believed in the existence of God. Nothing could be further from the truth—indeed, Einstein can be described more accurately as an outright atheist. Although his early upbringing was in a highly religious Jewish environment, he soon realized that many of the things described in the Old Testament were not consistent with physical laws. His great contributions to physics came from his belief in precise mathematical laws that govern the natural world. This rational approach is antithetical to the common religious notion of a supernatural God with powers that can overcome natural laws.

We can go as far as saying that, deep down, every person must have this rational streak. You cannot do good science if you do not believe in fundamental immutable laws that govern Nature. Tomorrow, if your computer breaks down, you know it is because some part of the system failed. You call a technician hoping she will find out what is wrong and fix it; you certainly don't pray to a God or go to a temple to get it fixed. It is interesting that we are born with this rational bent of mind—in fact, our very survival in the natural world depends on forming a rational picture of what we see, with no room for supernatural or magical events. Experiments have shown that human infants below the age of one—well before they are even able to talk—will get perturbed by magical events that do not conform to their rational model of the world (not falling when you go over the edge of a bed, for example). It is only later that we become mature enough to be able to enjoy magic shows by consciously suspending our rational belief during the magician's performance.

So what did Einstein really mean by the use of the word 'God' in his statement. Einstein of course believed in laws that govern Nature; so he saw the hand of God in the precise nature of physical laws, in their mathematical beauty and elegance, and in their simplicity. To him, the very fact that there were natural laws that the human mind could discover was evidence of a God, not a God who superseded these laws but one who created them. Thus his use of the word God is to be interpreted as the existence of natural laws of great mathematical beauty, whether they were already discovered or not.

B. Quantum mechanics versus general relativity

We now discuss the second part of Einstein's statement, about not playing dice. This relates to Einstein's reaction to the part of Nature described by *Quantum Mechanics*, which is undoubtedly one of the pillars of modern physics. He felt that natural laws could not be like the throw of dice, with inherent randomness or probability. But this is exactly what Quantum Mechanics tells us—that at the fundamental level Nature is inherently random, codified in Heisenberg's famous *Uncertainty Principle*. Thus, the second misunderstanding about Einstein's statement is that his opposition to Quantum Mechanics was the raving of an old man, a man well beyond his prime who did not understand the new physics. Well, we will see below why this is all a myth.

Einstein's great contributions to physics started in his *Annus Mirabilis*, the year 1905 whose centenary was celebrated as the World Year of Physics. In that year, Einstein published six seminal papers that revolutionized our understanding of the physical universe in three different directions. The papers dealt with:

1. The 'light-quantum' or the photon concept, and an explanation of the photoelectric effect.

2. The theory and explanation of Brownian motion.

3. The *Special Theory of Relativity*—a radically new view of space and time.

Einstein himself regarded only the first of these as truly revolutionary since it was the second major step (after Max Planck's work) in the development of quantum theory, whereas the Special Theory of Relativity belonged to the older classical theory. In addition, in the same year Einstein discovered the equivalence of mass and energy, encapsulated in perhaps the most famous equation of all

$$E = mc^2$$

Over the next decade (1905–15), Einstein used his understanding of the 'new' quantum hypothesis to make fundamental contributions to almost every area of physics where the idea could be applied—the specific heat of solids for example. But he was also quietly working on extending the Special Theory of Relativity to a more generalized theory encompassing a broader class of transformations between observers. He finally succeeded in 1915, when he published his *General Theory of Relativity*—a theory of unsurpassed beauty that explained gravitation as arising simply out of the geometry (or curvature) of spacetime. He also showed how the force of gravity had to go beyond the simple but successful theory of Newton. In particular, the force of gravity did not entail Newton's idea of instantaneous action-at-a-distance (e.g. the idea that the gravitational force of the Sun is felt instantaneously on Earth), but propagated at the speed of light as required by any relativistically correct theory. But this theory was entirely *classical*—there was nothing quantum about it.

The above summary of Einstein's contributions shows two important things about his work:

1. that he made fundamental contributions to our understanding of *quantum theory*, so that he, if anyone, was qualified to judge the nature of this theory, and

2. that, in formulating the General Theory of Relativity, he took our classical ideas beyond what anyone had done before.

Einstein spent the remainder of his life in a quest, albeit futile, for an even more generalized theory of relativity; a unified field theory that would geometrize all the forces of Nature and not just gravity. He was sure that the formalism of the General Theory of Relativity as propounded in 1915 was just a preliminary version that would be extended in due course of time. But he was unable to complete this in his lifetime. Here is what one of his biographers, Abraham Pais, has to say about his later work—*if Einstein had stopped working in 1915, the world of physics would not have lost much.*

So did Einstein's contributions to physics end in 1915; did he stop being a part of mainstream physics; was he so completely off the mark that his efforts were doomed before he even started? *Au contraire.* We will now see why his great mind chose to work on the alternate approach to physical theory based on General Relativity, and the basis for his lifelong opposition to Quantum Mechanics.

There were several unique and unprecedented features of General Relativity, three of which were particularly appealing to Einstein:

1. With the equations of General Relativity, Einstein found that space and time were no longer just a passive stage on which particles performed their acts, but were active members in the performance. Thus, the geometric structure of spacetime was determined by the matter in it, and of course the matter responded to this geometry and was constrained by its structure. The fact that space and time were now a part of the equations was unprecedented and its importance is beautifully expressed by Einstein himself — "*It is contrary to the mode of thinking in science to conceive of a thing ... which acts itself, but which cannot be acted upon.*"

2. This was the first theory in physics that was *nonlinear*. In other words, the gravitational field acted upon itself. An important consequence of this was that the equation of motion was contained in the field equations themselves. One did not have separate equations for the interactions between matter and for the response of matter to these interactions. By contrast, a linear theory like Maxwell's theory of electromagnetism could only describe the electromagnetic field interactions between charged matter. The response or inertial manifestation was contained in separate equations of motion given by Newton's three laws of motion that we learn in high school.

3. For the first time in physics, a theory predicted that the inertia of a body (its property of staying at rest or uniform motion unless acted upon by a force, contained in Newton's first law) depended on its surroundings. Much before Einstein, the philosopher-scientist Ernst Mach had the radical idea that perhaps the inertia of a body is a consequence of its interactions with the rest of the universe. In other words, the distant stars which define *inertial coordinates* of a system also determine its inertia of the system. The equations of General Relativity showed that the inertia of a system increases when it is placed in the vicinity of other heavy masses. Inertia was no longer some inherent 'God-given' property of a system, but was at least partly determined by the environment. Einstein's hope was that he would find a fully unified field theory which would show that all of the inertia—and not just part of it—was due to the interactions with the environment, *à la* Mach.

We are now in a position to understand Einstein's opposition to Quantum Mechanics. This was not the knee-jerk reaction of a person unqualified to speak about physics, but the considered opinion of an eminent scientist based on what he felt were several undesirable features of the theory. First, he was averse to the idea of randomness as a fundamental feature of any theory. He believed that randomness could appear as some form of statistical behavior but could not be a part of the law, just like a pack of cards that is shuffled according to deterministic laws still shows a random arrangement. But this was not its only undesirable feature. The theory was also inherently *nonlocal*, or had a kind of Newtonian action-at-a-distance built into it, but relativity had taught us that all interactions had to propagate at a finite speed. In a landmark paper published in 1935, Einstein highlighted the nonlocal and incomplete nature of Quantum Mechanics by proposing the famous Einstein–Podolsky–Rosen (EPR) paradox. The EPR paradox was a *gedanken* or thought experiment that brought out these undesirable features of Quantum Mechanics, and we all know that Einstein was a past master at thought experiments. Finally, of course, Einstein was opposed to the linear formalism of Quantum Mechanics as an approach to understanding Nature because we have seen above that only a nonlinear theory can contain the equations of motion.

There were thus three features of Quantum Mechanics that Einstein disapproved of—it was probabilistic, nonlocal, and linear. Despite this opposition, Einstein realized that it was a successful theory within its domain of applicability. He believed that a future unified field theory would have to reproduce the results of Quantum Mechanics, perhaps as a linear approximation to a deeper nonlinear theory. This was similar to how the relativistic gravitational field of General Relativity (with a finite propagation speed of the gravitational force) led to Newton's law of gravitation (with its action-at-a-distance force) in the non-relativistic limit. But Einstein was convinced that Quantum Mechanics was not the correct approach to deducing the fundamental laws of physics.

Unfortunately now, several decades after Einstein's death, the mainstream of physics does not take his approach seriously. One can only seek solace from the fact that Newton's theory of gravity was enormously successful until Einstein came along.

Chapter 4

Einstein as armchair detective—The case of stimulated radiation

T HIS chapter is devoted to the *modus operandi* of the genius Einstein. We look at his 1917 paper in which he predicted the existence of stimulated emission in radiative processes, well before the formalism of quantum mechanics was developed. In that sense, the paper belongs to the "old" quantum theory—but its prediction of stimulated emission is the key to understanding the operation of the laser, which finds myriad applications today.

A. Einstein's *modus operandi*

Einstein is rightly regarded as one of the greatest scientific geniuses of all time. Perhaps the most amazing and awe-inspiring feature of his work was that he was an "armchair scientist"—not a scientist who spent long hours in a laboratory conducting delicate experiments, but one who performed *gedanken* (thought) experiments while sitting in his favorite chair that nevertheless advanced our understanding of nature by leaps and bounds. Two of his greatest contributions are the special theory of relativity and the general theory of relativity, both abstract creations of his remarkable intellect. They stand out as scientific revolutions that completely changed our perceptions of nature—of space and time in the case of the special theory, and of gravity in the case of the general theory. It might be argued that the special theory of relativity was necessitated by experimental facts such as the constancy of the speed of light in different inertial frames, but the general theory was almost completely a product of Einstein's imagination. For a person to have achieved one revolution in his lifetime is great enough, but two revolutions seems quite supernatural.

But is it really so magical? While it is certain that Einstein was a one-of-a-kind genius, is it at least possible to understand the way in which his mind tackled these problems? I think the answer is yes, because deep inside Einstein was like a detective hot on a mystery trail, of course not one solving murder mysteries but one trying to unravel the mysteries of nature. Any keen follower of murder mysteries knows that there are two types of detectives—(i) those who get down on their hands and knees looking for some microscopic piece of clinching evidence at the scene of the crime, and (ii) "armchair detectives" who seem to arrive at the solution just by thinking logically about the various possibilities. Einstein was most certainly of the second kind, and true to this breed, he had his own *modus operandi*. In simple terms, his technique was to imagine nature in a situation where she contradicted established truths, and revealed new truths in the process. As a case in point, we will look at Einstein's 1917 paper titled "On the Quantum Theory of Radiation" where he predicted the existence of stimulated emission.* While Einstein will always be remembered for his revolutionary relativity theories, his contributions to

---

*English translation of the paper available in *Sources of Quantum Mechanics*, B. L. Van Der Waerden, ed. (Dover, New York, 1968).

the early quantum theory are certainly of the highest caliber, and the 1917 paper is a classic.

It is useful to first set the paper in its historical perspective. By the time Einstein wrote this paper, he had already finished most of his work on the relativity theories. He had earlier done his doctoral thesis on Brownian motion and was a pioneer of what is now called statistical mechanics. He was thus a master at using thermodynamic arguments. He was one of the earliest scientists to accept Planck's radiation law and its quantum hypothesis. He had already used it in 1905 for his explanation of the photoelectric effect. He was also aware of Bohr's theory of atomic spectra and Bohr's model of the atom, which gave some explanation for why atoms emitted radiation in discrete quanta. What he did **not know** in 1917 was any of the formalism of quantum mechanics, no Schrödinger equation and not the de Broglie hypothesis for the wave nature of particles that we learn in high school these days. Despite this, Einstein was successful in predicting many new things in this paper.

Let us now see what is Einstein's strategy in this paper. He is attempting to understand the interaction between atoms and radiation from a quantum mechanical perspective. For this, he imagines a situation where a gas of atoms is in thermal equilibrium with radiation at a temperature T. The temperature T determines both the Maxwell-Boltzmann velocity distribution of atoms, and the radiation density ρ at different frequencies through Planck's law. He assumes that there are two quantum states of the atom Z_n and Z_m, whose energies are ε_n and ε_m respectively, and which satisfy the inequality $\varepsilon_m > \varepsilon_n$. The relative occupancy W of these states at a temperature T depends on the Boltzmann factor as follows

$$W_n = p_n \exp(-\varepsilon_n/k_B T)$$

$$W_m = p_m \exp(-\varepsilon_m/k_B T)$$

(4.1)

where p_n is a number, independent of T and characteristic of the atom and its nth quantum state, called the degeneracy or "weight" of the particular state. Similarly, p_m is the weight of the mth state.

Einstein then makes the following basic hypotheses about the laws governing the absorption and emission of radiation.

1. Atoms in the upper state m make a transition to the lower state n by spontaneous emission. The probability dW that such a transition

occurs in the time dt is given by

$$dW = A_m^n dt$$

A_m^n in modern terminology is called the Einstein A coefficient. Since this process is intrinsic to the system and is not driven by the radiation field, it has no dependence on the radiation density.

2. Atoms in the lower state make a transition to the upper state by absorbing radiation. The probability that such a transition occurs in the time dt is given by

$$dW = B_n^m \rho dt \tag{4.2}$$

B_n^m is now called the Einstein B coefficient. The absorption process is driven by the radiation field, therefore the probability is directly proportional to the radiation density ρ at frequency ν.

3. The two postulates above seem quite reasonable. Now comes his new postulate, that there is a third process of radiative transition from the upper state to the lower state, namely stimulated emission, *driven by the radiation field*. By analogy with the probability for absorption, the probability for stimulated emission is

$$dW = B_m^n \rho dt \tag{4.3}$$

Einstein calls the processes in both 2 and 3 as "changes of state due to irradiation". We will see below how he is forced to include postulate 3 in order to maintain thermodynamic equilibrium.

The main requirement of thermodynamic equilibrium is that the occupancy of atomic levels given by Eq. (4.1) should not be disturbed by the absorption and emission processes postulated above. Therefore the number of absorption processes (type 2) per unit time from state n into state m should equal the number of emission processes (type 1 and 3 combined) out of state m into state n. This is called *detailed balance*. Since the number of processes from a given state occurring in a time dt is given by the occupancy of that state times the probability of a transition, the detailed balance condition is written as

$$p_n \exp(-\varepsilon_n/k_B T) B_n^m \rho = p_m \exp(-\varepsilon_m/k_B T)(B_m^n \rho + A_m^n) \tag{4.4}$$

Notice the importance of the third hypothesis about stimulated emission to make the equation consistent. If one does not put that in, the equation becomes

$$p_n \exp(-\varepsilon_n/k_B T) B_n^m \rho = p_m \exp(-\varepsilon_m/k_B T) A_m^n$$

which clearly will not work. At high temperatures, when the Boltzmann factor makes the occupancy of the two levels almost equal, the rate of absorption on the LHS increases as the radiation density increases. But the rate of emission on the RHS does not increase because spontaneous emission is independent of the radiation density. Thermodynamic equilibrium will therefore not be maintained. This is vintage Einstein—he imagines a situation that forces a contradiction with what he "knows", namely thermal equilibrium, and uses it to obtain a new result, namely stimulated emission during radiative transfer.

With the grace and confidence of an Olympic hurdler, Einstein now moves on to make quantitative predictions based on the bold new hypothesis. He first uses the high temperature limit to derive a relation between the coefficients for absorption and stimulated emission. Under the reasonable assumption that $\rho \to \infty$ as $T \to \infty$, the spontaneous emission term on the RHS of Eq. (4.4) can be neglected at high temperatures. From this, it follows that

$$p_n B_n^m = p_m B_m^n$$

By substituting this result in Eq. (4.4), Einstein obtains a new, simple derivation of Planck's law

$$\rho = \frac{A_m^n/B_m^n}{\exp[(\varepsilon_m - \varepsilon_n)/k_B T] - 1}$$

Notice that he will not get the correct form of this law if he did not have the stimulated emission term in Eq. (4.4). Another reason for him to be confident that his three hypotheses about absorption and emission are correct. He then compares the above expression for ρ with Wien's displacement law

$$\rho = v^3 f(v/T)$$

to obtain

$$\frac{A_m^n}{B_m^n} = \alpha v^3 \qquad \text{and} \qquad \varepsilon_m - \varepsilon_n = hv$$

with constants α and h. The second result is well known from the Bohr theory of atomic spectra. Einstein is now completely sure that his three hypotheses about radiation transfer are correct since he has been able to derive both Planck's law and Bohr's principle based on these hypotheses.

Einstein does not stop here. He now considers how interaction with radiation affects the atomic motion in order to see if he can predict new features of the momentum transferred by radiation. Earlier he had argued that thermal equilibrium demands that the occupancy of the states remain undisturbed by interaction with radiation. Now he argues that the Maxwell-Boltzmann velocity distribution of the atoms should not be disturbed by the interaction. In other words, the momentum transfer during absorption and emission should result in the same statistical distribution of velocities as obtained from collisions. From kinetic theory, we know that the Maxwell-Boltzmann velocity distribution results in an average kinetic energy along each direction given by

$$\frac{1}{2}M\langle v^2 \rangle = \frac{1}{2}k_B T \tag{4.5}$$

This result should remain unchanged by the interaction with radiation.

To calculate the momentum change during radiative transfer, Einstein brings into play his tremendous insight into Brownian motion. As is now well known from the Langevin equation, he argues that the momentum of the atom undergoes two types of changes during a short time interval τ. One is a frictional or damping force arising from the radiation pressure that systematically opposes the motion. The second is a fluctuating term arising from the random nature of the absorption-emission process. It is well known from Brownian motion theory that the atoms would come to rest from the damping force if the fluctuating term were not present. Thus, if the initial momentum of the atom is $p_i = Mv$, then after a time τ, the momentum will have the value

$$p_f = Mv - Rv\tau + \Delta$$

where the second term is the damping term and the last term is the fluctuating term. If the velocity distribution of the atoms at temperature T is to remain unchanged by this momentum transfer process, the average of the above quantity must be equal to Mv, and the mean values of the squares of these quantities—which determines the energy distribution—must also be equal

$$\langle (Mv - Rv\tau + \Delta)^2 \rangle = \langle (Mv)^2 \rangle$$

Since we are only interested in the systematic effect of v on the momentum change due to interaction with radiation, v and Δ can be regarded as independent statistical processes and the average of the cross term $v\Delta$ can be neglected. This yields

$$\langle \Delta^2 \rangle = 2RM \langle v^2 \rangle \tau$$

To maintain consistency with kinetic theory, the value of $\langle v^2 \rangle$ in the above equation must be the same as the one obtained from Eq. (4.5). Thus

$$\frac{\langle \Delta^2 \rangle}{\tau} = 2Rk_BT \tag{4.6}$$

This is the equation that will tell Einstein if his hypotheses about momentum transfer are correct. In other words, he assumes that the radiation density is given by Planck's law, and calculates R and $\langle \Delta^2 \rangle$ based on some hypotheses about momentum transfer during radiative processes. If the hypotheses are valid, the above equation should be satisfied identically in order not to contradict thermal equilibrium.

His main hypothesis about momentum transfer is that, if the photon behaves like a localized packet of energy E, it must also carry directional momentum of E/c. Without going into mathematical details, we just outline his approach for calculating R and $\langle \Delta^2 \rangle$.

For calculating R, he uses the following argument. In the laboratory frame in which the atom has a velocity v, the radiation is isotropic. But in the rest frame of the atom, the radiation is anisotropic because of the Doppler shift. This gives rise to a velocity-dependent radiation density and a velocity-dependent probability of absorption and stimulated emission [from Eqs. (4.2) and (4.3)]. The average momentum transferred to the atom is calculated from the modified rates of absorption stimulated emission, thus yielding R. R does not depend on the rate of spontaneous emission because spontaneous emission occurs independently of the radiation field, and is therefore isotropic in the rest frame of the atom.

Calculating $\langle \Delta^2 \rangle$ is relatively simpler. If each absorption or emission process gives a momentum kick of E/c in a random direction, the mean squared momentum after ℓ kicks is simply $\ell \times (E/c)^2$. ℓ is equal to twice the number of absorption processes taking place in the time τ since each absorption process is followed by an emission process.

Using this approach, Einstein calculates R and $\langle \Delta^2 \rangle$. He shows that Eq. (4.6) is satisfied identically when these values are substituted, which implies that the velocity distribution from kinetic theory is not disturbed if and only if momentum exchange with radiation occurs in units of E/c in a definite direction.

He thus concludes the paper with the following observations. There must be three processes for radiative transfer, namely absorption, spontaneous emission, and *stimulated emission*. Each of these interactions is quantized and takes place by interaction with a single radiation bundle. The radiation bundle (which we today call a photon) carries not only energy of $h\nu$ but also momentum of $h\nu/c$ in a well-defined direction. The momentum transferred to the atom is in the direction of propagation for absorption and in the opposite direction for emission. And finally, ever loyal to his dislike for randomness in physical laws ("God does not play dice ...!"), he concludes that one weakness of the theory is that it leaves the duration and direction of the spontaneous emission process to "chance". However, he is quick to point out that the results obtained are still reliable and the randomness is only a defect of the "present state of the theory".

What far reaching conclusions starting from an analysis of simple thermodynamic equilibrium. This is a truly great paper in which we see two totally new predictions. First, he predicts the existence of stimulated emission. And to top that, for the first time, he shows that each light quantum carries discrete momentum, in addition to discrete energy. He shows that the directional momentum is present even in the case of spontaneous emission. Thus an atom cannot decay by emitting "outgoing radiation in the form of spherical waves" with no momentum recoil.

Today his conclusions about momentum transfer during absorption and emission of radiation have been abundantly verified. Equally well verified is his prediction of stimulated emission of radiation. Stimulated emission is the mechanism responsible for operation of the laser, which is used in everything from home computers and CD players to long-distance communication systems. Stimulated emission, or more correctly stimulated scattering, underlies our understanding of the phenomenon of

Bose-Einstein condensation. It plays an important role in the explanation of superconductivity and superfluidity. The two predictions, momentum transfer from photons and stimulated emission, play a fundamental role in an important area of research, namely laser cooling of atoms. In laser cooling, momentum transfer from laser photons is used to cool atoms to very low temperatures of a few millionths of a degree above absolute zero. Perhaps fittingly, it is the randomness or "chance" associated with the spontaneous emission process (which he disliked so much) that is responsible for the entropy loss associated with cooling. In other words, as the randomness from the atomic motion gets reduced by cooling, it gets added to the randomness in the radiation field through the spontaneous emission process, thus maintaining consistency with the second law of thermodynamics.

B. Discussion

We have thus seen how Einstein was able to use the principle of thermodynamic equilibrium to imagine a situation where radiation and matter were in dynamical equilibrium and from that predict new features of the radiative transfer process. As mentioned before, this was a recurring theme in his work, a kind of *modus operandi* for the great "detective". In his later writings, he said that he always sought one fundamental governing principle from which he could derive results through these kind of arguments. He found such a principle for thermodynamics, namely the second law of thermodynamics, which states that *it is impossible to build a perpetual motion machine*. He showed that the second law was a necessary and sufficient condition for deriving all the results of thermodynamics. His quest in the last four decades of his life was to geometrize all forces of nature. In this quest, he felt that he had indeed found the one principle that would allow him to do this uniquely, and this was the *principle of relativity*

> *the laws of physics must look the same to all observers no matter what their state of motion.*

He had already used this principle to geometrize gravity in the general theory of relativity. His attempts at geometrizing the electromagnetic force—the only other long-range force, while the strong and weak forces are short range—remained an unfulfilled dream.

C. Examples of *gedanken* experiments

We present two examples of *gedanken* experiments that illustrate the Einstein technique for arriving at new results. Both of these experiments yield results associated with the general theory of relativity, but are so simple and elegant that they can be understood without any knowledge of the complex mathematical apparatus needed for the general theory. The first experiment is due to Einstein himself, while the second is due to Hermann Bondi.

1. Need for curved spacetime for gravity

This is a thought experiment devised by Einstein to arrive at the conclusion that the general theory of relativity is an extension of the special theory which requires curved spacetime, or spacetime in which the rules of plane (Euclidean) geometry do not apply. The "known" facts are the results of special theory of relativity applicable to inertial systems, and the equivalence principle which states that inertial mass is exactly equal to gravitational mass. Einstein's argument proceeds as follows.

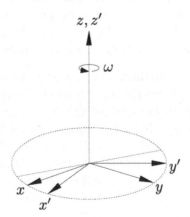

Figure 4.1: Coordinate systems with relative rotation between them.

Imagine two observers or coordinate systems O and O′. Let the z' axis of O′ coincide with the z axis of O, and let the system O′ rotate about the z axis of O with a constant angular velocity, as shown in Fig. 4.1. Thus O is an inertial system where the laws of special relativity apply, while O′ is a non-inertial system. Imagine a circle drawn about the origin in the $x'y'$ plane of O′ with some given diameter. Imagine, further, that we have a large number of rigid rods, all identical to each other. We lay these rods in series along the circumference and the diameter of the circle, at rest with respect to O′. If the number of rods along the circumference is U and the number of rods along the diameter is D, then, if O′ does not rotate with respect to O, we have (from plane geometry)

$$\frac{U}{D} = \pi$$

However, if O′ rotates, we get a different result. We know from special relativity that, relative to O, the rods on the circumference undergo Lorentz contraction while the rods along the diameter do not undergo this contraction (the relative motion is perpendicular to the diameter). Therefore, we are led to the unavoidable conclusion that

$$\frac{U}{D} > \pi$$

i.e. the laws of configuration of rigid bodies with respect to O′ are not in accordance with plane geometry. If, further, we place two identical clocks, at rest with respect to O′, one at the periphery and one at the center of the circle, then with respect to O the clock at the periphery will go slower than the clock at the center from special relativity. A similar conclusion will be reached by O′, i.e. the two clocks go at different rates.

We thus see that space and time cannot be defined with respect to O′ as they were defined in special theory of relativity for inertial systems. But, according to the equivalence principle, O′ can also be considered a system at rest with respect to which there is a gravitational field (corresponding to the centrifugal force field and the Coriolis force field). We therefore arrive at the following remarkable result—the gravitational field influences and even determines the geometry of the spacetime continuum, and this geometry is not Euclidean. From this conclusion, Einstein goes on to develop a curved spacetime theory of gravitation.

2. Gravitational redshift

This example illustrates the use of a thought experiment to calculate the difference in rates between two clocks placed at different gravitational potentials, called the gravitational redshift. We have already seen in the first example how the rate of the clock at the periphery differs from the rate of the clock at the center. Here, we derive a quantitative value for this difference using an Einstein-like *gedanken* experiment, first conceived by Bondi. The "known" things are the second law of thermodynamics and the special relativistic energy-mass relationship

$$E = mc^2$$

Imagine a series of buckets on a frictionless pulley system, as shown in Fig. 4.2. Each bucket contains an atom capable of absorbing or emitting a photon of energy $h\nu$. The system is in a uniform gravitational field with acceleration g. If the photon frequency were unaffected by the gravitational field, we can operate the system as a perpetual motion machine in the following way. Imagine that the pulleys rotate clockwise and that all the atoms on the left are in the ground state and the atoms on the right are in the excited state. The lifetime in the excited state is such that, on average, every time a bucket reaches the bottom the atom inside decays to the ground state and emits a photon. Suitable reflectors direct this photon to the corresponding bucket at the top so that the atom inside absorbs the photon and goes into the excited state. All the excited state atoms on the right have more energy and, from the relation $E = mc^2$, are therefore heavier by an amount $\Delta m = h\nu/c^2$. The heavier masses are accelerated down by the gravitational field and the system remains in perpetual motion. The excess gravitational potential energy can be converted to unlimited useful work, in violation of the second law of thermodynamics.

The solution to the paradox lies in postulating that the frequency of the photon emitted by the atom at the bottom is not the same as the frequency of the photon when it reaches the top. Let the two frequencies be ν and ν' respectively. Then the additional mass for the atom at the top by absorbing a photon of frequency ν' is $h\nu'/c^2$, and the potential energy of this excess mass at a height H between the two buckets is $h\nu'/c^2 \times gH$. To maintain consistency with the second law of thermodynamics, this excess energy should exactly compensate for the loss in energy of the photon as its

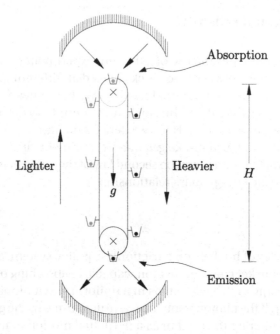

Figure 4.2: Bondi's perpetual motion machine. The buckets on the right side contain atoms that have higher energy and are thus heavier than the atoms on the left side. When a bucket reaches the bottom, the atom inside emits a photon which is absorbed by the corresponding atom in the top bucket. The heavier buckets on the right keep falling down in the gravitational field and their gravitational energy can be converted to useful work *in perpetuum*. The resolution to the paradox is that the photon absorbed at the top has a lower frequency (and hence smaller energy) than the photon emitted at the bottom.

frequency changes from v to v'

$$\frac{hv'}{c^2} gH = h(v - v')$$

which yields

$$\frac{v' - v}{v'} = -\frac{gH}{c^2}$$

i.e. the relative frequency shift is given by gH/c^2 and is negative (redshift) at the location where the gravitational potential is higher. The shift can be understood from the fact that the photon is also affected by the gravitational field and therefore loses energy as it climbs up the potential. Since the photon always travels at the speed c, it loses energy by changing its frequency. This result explains why, in the first example, the clock at the center goes slower than the clock in the periphery according to O'. With respect to O', there is a gravitational field (corresponding to the centrifugal force) pointing away from the center. The clock at the center is at a higher gravitational potential and hence goes slower.

The gravitational redshift on the surface of the earth is very tiny at any reasonable height, but it was experimentally verified in a remarkable experiment by Pound and Rebka in 1960.[*] They measured the frequency shift of a recoilless Mössbauer transition between the top and bottom of a building at Harvard University—a height difference of about 25 m. The relative frequency shift measured was a tiny 3 parts in 10^{14}, consistent with the above calculation!

[*]R. V. Pound and J. G. V. Rebka, "Apparent weight of photons," Phys. Rev. Lett. **4**, 337–341 (1960).

Chapter 5

Standards for units

COMMUNITY living of humans has necessitated the adoption of a common set of measures for quantities such as mass, length, and time. While settlements with little or no trade with others could afford to have their own set of measures, inter-community trade and commerce has made it necessary for people to adopt common ones. This requirement is all the more important in modern science, because the results of experiments in different labs can be compared only if all scientists agree on a common set of measures. The scientific study of such measures is called *metrology*. In most countries, standards for units are maintained by national institutions, in coordination with similar bodies in other countries. In this chapter, we will see the evolution of standards for commonly used units from early times to their modern "democratic" definitions.

The modern set of units accepted by all countries is the SI system—abbreviated from French *Le Système International d'Unités*. The standards for them have to satisfy several important requirements:

1. They should be *invariant*, i.e. the definition should not change with time.

2. They should be *reproducible*, i.e. it should be possible to make accurate and faithful copies of the original.

3. They should be on a *human scale*, i.e. of a size useful for everyday purposes.

4. They should be *consistent* with physical laws, i.e. the minimum number of independent units should be defined and other units derived from these using known physical laws.

An important way to achieve these goals is to define standards based on fundamental constants of nature. This helps us get away from artifacts and makes the definition *democratic*.

A. Time standards

Time keeping is as old as the earth itself since all it requires is a periodic process. The rotation and revolution of the earth give rise to daily and yearly cycles, respectively, and this has been used by nature as a clock much before humans evolved. Most living organisms show diurnal (24 hour) rhythms regulated by the sun, and seasonal patterns that repeat annually. It is therefore natural that the earliest man-made clocks also relied on the sun. One example is the sundial. It consisted of a pointer and a calibrated plate, on which the pointer cast a moving shadow. Of course this worked only when the sun was shining. The need to tell the time even when the sun was not out, such as on an overcast day or in the night, caused man to invent other clocks. In ancient Egypt, water clocks known as *clepsydras* were used. These were stone vessels with sloping sides that allowed water to drip at a constant rate from a small hole near the bottom. Markings on the inside surface indicated the passage of time as the water level reached them. While sundials were used during the day to divide the period from sunrise to sunset into 12 equal hours, water clocks were used from sunset to sunrise. However, this definition made the "hours" of the day and night different, and also varying with the season as the length of the day changed. Other more accurate clocks were used for measuring small intervals of time. Examples included candles marked in increments, oil lamps with marked reservoirs, hourglasses filled with sand, and small stone or metal mazes filled with incense that would burn at a certain rate.

Time measurements became significantly more accurate only with the use of the pendulum clock in the 17th century. Galileo had studied the pendulum as early as 1582, but the first pendulum clock was built by Huygens only in 1656. As discussed in "Oscillations" chapter, the time period of the pendulum clock is $T = 2\pi\sqrt{\ell/g}$, and therefore depends only on the length of the pendulum and the local value of the acceleration due to gravity. Huygens' clock had an unprecedented error of less than 1 minute per day. Later refinements allowed him to reduce it to less than 10 seconds a day. While very accurate compared to previous clocks, pendulum clocks still showed significant variations because even a few degrees change in the ambient temperature could change the length of the pendulum due to thermal expansion. Therefore clever schemes were developed in the 18th and 19th centuries to keep the time period constant during the course of the year by compensating for seasonal changes in length. Pendulum clocks were also not reproducible from one place to another because of variations in the value of g. Finally, this and all other mechanical clocks—based on oscillations of a balance wheel as in wrist watches, for example—suffered from unpredictable changes in time-keeping accuracy with wear and tear of the mechanical parts.

Mechanical time keeping devices were useful in telling the time whenever we wanted. As a time standard, however, the rotation rate of the earth proved to be more regular than anything man made. Despite the many advances in technology of mechanical clocks, they were still less accurate than the "earth day". Moreover, the success of astronomical calculations in the 19th century led scientists to believe that any irregularities in the earth's rotation rate could be adequately accounted for by theory. Therefore, until 1960, the unit of time—a second—was defined as the fraction 1/86400 of the "mean solar day" as determined by astronomical theories. The earth's rotation rate was the primary time standard, while mechanical clocks were used as secondary standards whose accuracy was determined by how well they kept time with respect to earth's rotation.

As clock accuracies improved in the first half of the 20th century, especially with the development of the quartz crystal oscillator, precise measurements showed that irregularities in the rotation of the earth could not be accounted for by theory. In order to define the unit of time more precisely, in 1960 a definition given by the International Astronomical Union based on the tropical year was adopted. However, scientists were still looking for a truly universal standard based on some physical constant. They were able to do this in the late 1960s based on the predictions of quantum mechanics. Planck's famous relation between the energy and frequency of a photon $E = h\nu$, which signaled the birth of quantum mechanics, implies that atoms have a unique internal frequency corresponding to any two energy levels. Since these energy levels are characteristic of an atom anywhere in the universe, a definition based on the atom's internal frequency would be truly universal. Once the ability to measure such frequencies accurately was developed, the following definition of the unit of time was adopted:

> *the second is the duration of 9192631770 periods of the radiation corresponding to the transition between the two hyperfine levels of the ground state of the cesium-133 atom.*

Modern atomic clocks are built according to the above definition. The clock consists of a vacuum chamber with a cesium atomic beam, and a radio-frequency oscillator which is tuned to drive the atoms between the two hyperfine levels. There is maximum transfer of energy from the laboratory oscillator to the atoms when the atoms are in **resonance**, i.e. the oscillator frequency matches the internal frequency of the atom. A feedback circuit locks these two frequencies and ensures that the laboratory clock does not drift with respect to the atom frequency. With this feedback, the best atomic clocks (used as primary standards) are so precise that they lose less than 1 second in a million years! The time standard is also universal in the sense that anybody who builds an atomic clock will get exactly the same frequency because all atoms are identical, and their behavior under controlled experimental conditions is the same anywhere in the universe.

Life in the modern high-technology world has become crucially dependent on precise time. Computers, manufacturing plants, electric power grids, satellite communication, all depend on ultra-precise timing. One example that will highlight this requirement is the Global Positioning System (GPS), which uses a grid of satellites to tell the precise location of a receiver anywhere on earth. Transport ships plying the vast oceans of the world now almost completely rely on the GPS system for navigation. The system works by triangulating with respect to the three nearest satellites. The distance to each satellite is determined by timing the arrival of pulses traveling at the speed of light. It takes a few millionths of a second for the signal to reach the receiver, which gives an idea of how precise the timing has to be to get a differential reading between the three signals. Such demands of modern technology are constantly driving our need for ever more precise clocks. The cesium fountain clock at National Institute of Standards and Technology (NIST) in the United States, and other national labs, is one example of how the latest scientific research in laser cooling has been used to improve the accuracy of the clock by a factor of 10. Another promising technique is to use a single laser-cooled ion in a trap, and define the second in terms of an optical transition of the ion. The ion trap represents an almost ideal environment where the ion can be held for a long time.

B. Length standards

For a long time, length standards were based on arbitrary measures such as the length of the arm or foot of a tribal chief. The first step in the definition of a rational measure for length was the definition of the "meter" by the French in 1793, as a substitute for the yard. It was defined such that the distance between the equator and the north pole, as measured on the great circle through Paris, would be 10,000 kilometers. This gave a convenient length scale of one meter, which was very close to the old yard *but now invariant.* A prototype meter scale made of a platinum-iridium bar was kept in Paris. The alloy was chosen for its stability and exceptionally low thermal expansivity. Copies of this scale were sent to other nations and were periodically recalibrated by comparison with the prototype.

The limitations of the artifact meter scale in terms of invariance became apparent as more precise experiments started to be conducted in the 20th century. As for time standards, the results of quantum mechanics provided a solution. Each photon has not only a frequency v but also a wavelength λ, with the two related by the speed of light c

$$v\lambda = c$$

If an atom is excited, it decays to the ground state by emitting a photon of well defined wavelength corresponding to the energy difference between the two levels. These photons form a unique line spectrum, or wavelength signature, of the particular atom. The wavelength of photons can be measured precisely in optical interferometers by counting fringes as a function of path length difference in the two arms. Each fringe corresponds to a path length difference of λ. Therefore, in 1960, the artifact meter scale was replaced by a definition based on the wavelength of light:

> *1 meter is equal to* 1650763.73 *wavelengths of the orange-red line in the radiation spectrum from electrically excited krypton-86 atoms.*

Anybody who wanted to make a standard meter scale could do it by comparing to the krypton line, thus making it universally reproducible.

The advent of lasers has made the measurement of wavelength in optical interferometers very precise. In addition, the frequency of the laser can be measured accurately with respect to atomic clocks. In order to eliminate the fact that the definition of the meter was tied to the wavelength of a particular line from the krypton atom, the meter was redefined in 1983 using the above frequency-wavelength relation:

1 meter is the length of the path traveled by light in vacuum during a time interval of 1/299792458 of a second.

It is important to note that this definition makes the speed of light exactly equal to 299792458 m/s and demonstrates our faith in the special theory of relativity, which postulates that the speed of light in vacuum is a constant. As our ability to measure the meter becomes more accurate, it is the definition of the meter that will change in order to keep the numerical value of the speed of light a constant. In this sense, we have actually dispensed with an independent standard for length and made it a derived standard based on the standard of time and a fundamental constant of nature c.

Using iodine-stabilized HeNe lasers, the wavelength of light and thus the definition of the meter is now reproducible to about 2.5 parts in 10^{11}. In other words, if we were to build two meter scales based on this definition, their difference would be about one million times smaller than the thickness of a human hair.

C. Mass standards

In olden societies, mass standards were based on artifacts such as the weight of shells or of kernels of grain. The first scientific definition of mass adopted in the 18th century was the "gram", defined as the mass of 1 cubic centimeter of pure water at 4°C. This definition made the density of water exactly 1 g/cc. The definition survived till almost the end of the 19th century. It was replaced in 1889 when the 1st General Conference on Weights and Measures sanctioned an international prototype kilogram to be made of a cylinder of platinum-iridium alloy and kept at the International Bureau of Weights and Measures in France. It was declared that henceforth the prototype would be the unit of mass.

Among the base units, mass is the only one that is still based on an artifact and not on some fundamental property of nature. Environmental contamination and material loss from surface cleaning are causing the true mass of the kilogram to vary by about 5 parts in 10^8 per century relative to copies of the prototype in other nations. There are many physical constants that depend on mass, and the drift of the mass standard means that these constants have to be periodically revised to maintain consistency within the SI system.

There are several proposals to replace the mass standard with a universal one based on physical constants. There are two major approaches taken in this effort—one based on electrical measurements, and the other based on counting atoms. In the electrical measurement approach, an electrical force is balanced against the gravitational force on a kilogram mass. The electrical force is determined by the current and voltage used to produce the force. As we will see later, electrical standards are now based on fundamental constants, therefore the kilogram will also be related to fundamental constants.

The atom counting approach is a proposal that uses the tremendous advances made in silicon processing technology in recent decades. It is now possible to make large single crystals of silicon with very high purity and a defect rate that is less than one part in 10^{10}, i.e. atoms in the crystal are stacked perfectly and less than one atom in 10 billion is out of place! Such single crystals also cleave along certain symmetry planes with atomic precision, i.e. the crystal facet after a cut is atomically flat. Laser interferometry can be used to measure the distance between the outer facets of the crystal very precisely. This yields the precise volume of the crystal. Similarly, X-ray diffraction can be used to measure the spacing between successive atoms (lattice spacing) very precisely. The volume and the lattice spacing effectively tell us how many atoms there are in the crystal. From the definition of a mole we know that N_A silicon atoms will have a weight of $M_{Si} \times 10^{-3}$ kilograms, where N_A is Avogadro's constant and M_{Si} is the atomic mass of silicon. Therefore, a knowledge of the physical constants N_A and M_{Si}, combined with length measurements for the size of the crystal, will give the mass of the sample in kilograms. The present limitation in using a silicon mass standard is the precise knowledge of N_A. However, there are several experiments currently being performed that might yield a more precise and useful value for this constant. After this happens, the kilogram would be redefined as being the mass of a specific number of silicon atoms, and we would have eliminated the only remaining artifact standard.

D. Electrical standards

We next discuss the use of fundamental constants in defining electrical standards. Electrical units can all be related to the base units of mass, length, and time through physical laws. For example, Coulomb's law for the force between two charges q_1 and q_2 separated by a distance r is expressed as

$$F = K\frac{q_1 q_2}{r^2}$$

The proportionality constant K appearing in the above equation can be interpreted in two ways. From a physicist's point of view, it is just a matter of definition and can be set to 1 without any change in the underlying physics. In such a case, the units for measuring charges are so defined that q^2/r^2 has the dimensions of force. Thus charge becomes a derived unit which can be related to the dimensions of mass, length, and time through the above equation. From a practical point of view (which is followed in the SI system of units), it is useful to assign an independent unit for charge (coulomb in the SI system).* The constant K then serves to match the dimensions on both sides of the equation. Thus in the SI system $K = 1/4\pi\epsilon_0$ with ϵ_0 having units of farad/meter. It should be emphasized that both points of view are valid; however the latter introduces concepts such as permittivity of vacuum which needlessly complicates our understanding of the physics.

Whatever the point of view, the units have to be consistent with Coulomb's law. Therefore, two unit charges placed unit distance apart in vacuum should experience a force of K units. This is how electrical standards have to be finally related to the mechanical standards in a consistent manner. While it may not be practical to measure forces on unit charges this way, other consequences of the electromagnetic laws make a practical comparison of electrical and mechanical units possible. For example, it is possible to measure electrical forces in current carrying conductors by balancing them against mechanical forces. This would be a realization of the ampere.

These traditional methods of defining electrical standards showed a lot

---

*In SI units, it is actually the ampere that is the base unit; it is defined as the constant current that has to flow along two infinite parallel conductors placed 1 meter apart so as to produce a force of 2×10^{-7} N/m. In this system, 1 coulomb is 1 ampere-second.

of variation over time. However, recently it has been possible to base the definitions of electrical quantities using invariant fundamental constants. Two effects are used for their definition—the Josephson effect and the quantum Hall effect. The Josephson effect relates the frequency of an AC current generated when a DC voltage is applied across a tunnel junction between two superconductors. The frequency of the current is given by

$$f = \frac{2e}{h}V$$

where e is the charge on the electron. In SI units, a DC voltage of 1 μV produces an AC current with a frequency of 483.6 MHz. The DC voltage can therefore be related to the time or frequency standard through the fundamental constant $h/2e$. The *quantum Hall effect* is a phenomenon discovered by von Klitzing in 1980. He showed that at low temperatures and high magnetic fields, the Hall resistance in certain semiconductor samples shows quantized steps. The fundamental unit of resistance is h/e^2 (about 25.7 kΩ in SI units), and the steps occur at values of this constant divided by an integer i. Since 1990, the SI standard of ohm is defined using the $i = 4$ step in the Hall resistance of a semiconductor sample. Given the robust nature of the steps, it is easy to reproduce the resistance and its value is determined only by physical constants. The standard volt and ohm in the SI system have thus been successfully tied to fundamental constants of nature. The definition of current (ampere) follows from Ohm's law

$$I = V/R$$

E. Discussion

We have thus seen a trend where fundamental constants play an increasingly important role in eliminating artifact standards and in deriving some units from others. Our faith in physical laws makes us believe that these constants are truly constant, and do not vary with time. Ultimately, we would like to find enough fundamental constants that we are really left with only one defined standard and all others are derived from it—just like we have been able to do for length and time by specifying the value of c, so that both are tied to one Cs time standard. All indications are that when we understand the force of gravity from a quantum mechanical perspective, mass (or inertia) and spacetime will be linked in a fundamental way. When this happens, the unit of mass will most likely be fundamentally related to the unit of time. We will be left with just one arbitrary definition for "second" which sets the scale for expressing all the laws of physics. Someday, if we were to meet an intelligent alien civilization and would like to compare their scientific knowledge with ours, we would only need to translate their time standard to ours.

F. Additional items

1. A brief history of time-keeping

The earliest man made clocks in the world were sundials. They have evolved from simple designs of flat horizontal or vertical plates to more elaborate ones that compensate for the motion of the sun in the sky during the course of the year. The oldest clocks that did not rely on observation of celestial bodies were water clocks. They were designed to either drip water from a small hole or fill up at a steady rate. Elaborate mechanical accessories were added to regulate the rate of flow of water and display the time. But the inherent difficulty in controlling the flow of water led to other approaches for time keeping.

The first mechanical clocks appeared in 14th century Italy, but they were not significantly more accurate. Accuracy improved only when the Dutch scientist Huygens made the first pendulum clock in 1656. Around 1675, Huygens also developed the balance wheel and spring assembly, which is still found in mechanical wrist watches today. In the early 18th century, temperature compensation in pendulum clocks made them accurate to better than 1 second per day.

John Harrison, a carpenter and self-taught clockmaker, refined temperature compensation techniques and added new methods of reducing friction. He constructed many "marine chronometers"—highly accurate clocks that were used on ships to tell the time from the start of the voyage. A comparison of local noon—the time at which the sun was at its highest point—with the time reading on the clock—which would give the time of the noon at the starting point—could be used to determine the longitude of the ship's current position. The British government had instituted the *Longitude Prize* so that ships could navigate on transatlantic voyages without getting lost; specifically, the prize was for a clock capable of determining longitude to within half a degree at the end of a voyage from England to Jamaica. Harrison's prize-winning design (in 1761) kept time on board a rolling ship nearly as well as a pendulum clock on land, and was only 5.1 seconds slow after 81 days of rough sailing, about 10 times better than required.

The next improvement was the development of the nearly free pendulum at the end of the 19th century with an accuracy of one hundredth of a second per day. A very accurate free-pendulum clock called the Shortt clock, was demonstrated in 1921. It consisted of two pendulums, one a slave and the other a master. The slave pendulum gave the master pendulum gentle pushes needed to maintain its motion, and also drove the hands of the clock. This allowed the master pendulum to remain free from mechanical tasks that would disturb its regularity.

Time keeping was revolutionized by the development of quartz crystal clocks in the 1920s and 30s. Quartz is a piezoelectric material, meaning that it generates an electric field when mechanical stress is applied, and changes shape when an electric field is applied. By cutting the crystal suitably and applying an electric field, the crystal can be made to vibrate like a tuning fork at a constant frequency. The vibration produces a periodic electrical signal that can be used to operate an electronic display. Quartz crystal clocks are far superior than mechanical clocks because they have no moving parts to disturb their regular frequency. They dominate the commercial market due to their phenomenal accuracy and low cost. For less than a dollar, you can get a watch that is accurate to about 1 second per year. And you do not have to worry about winding the clock every day or replacing the battery more than once a year or so.

Despite this success, quartz crystal clocks ultimately rely on a mechanical vibration whose frequency depends critically on the crystal's size and shape. Thus, no two crystals can be precisely alike and have exactly the same frequency. That is why they cannot be used as primary time standards. And atomic clocks, which *are* used as primary standards, are exactly reproducible (when operated under the prescribed conditions).

In Fig. 5.1, we see that the accuracy of clocks in the last millennium has increased exponentially *on a logarithmic scale*. Extrapolating into the near future, we expect *optical* atomic clocks to be 1000 times more accurate than the current radio-frequency standard.

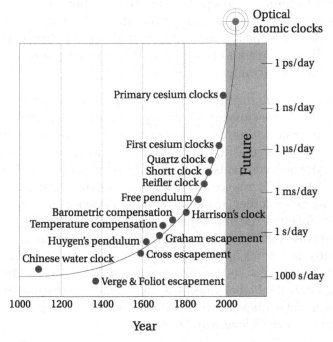

Figure 5.1: Exponential improvement in clock accuracy over the last millennium. Note the logarithmic scale of the y axis.

2. Dimensional analysis

It is somewhat instructive to play with dimensions of fundamental constants to see how they result in phenomena that can be used for defining standards. For electrical phenomena at the quantum level, there are two constants—the charge on an electron e and Planck's constant h. In the SI system, e has units of coulombs (C), and h has units of joules/hertz (J/Hz). It can be verified that h/e has the dimensions of volts/hertz (V/Hz), which is the quantum mechanical voltage-to- frequency conversion factor. The actual factor in the Josephson effect is $h/2e$, but the factor of 2 can be understood when we remember that the basic charge carriers in a superconductor are paired electrons, which thus carry charge of $2e$. Similarly h/e^2 has dimensions of ohms (Ω), and is the basic unit of quantum resistance in a semiconductor sample.

Continuing with this theme, we expect a deeper understanding of inertia when we develop a satisfactory theory of relativistic quantum gravity. The fundamental constants that are expected to be important in such a theory are c (for relativistic), h (for quantum), and Newton's constant G (for gravity). These constants have values 3×10^8 m/s, 6.6×10^{-34} J/Hz, and 6.7×10^{-11} Nm^2/kg^2, respectively. The three constants can be combined in various ways to get other constants that have dimensions of mass, length, and time. Together the new constants set the scale at which we expect quantum gravity effects to become significant. This is called the Planck scale. Thus the Planck length $\sqrt{hG/c^3}$ is 1.6×10^{-35} m, the Planck time $\sqrt{hG/c^5}$ is 5.4×10^{-44} s, and the Planck mass $\sqrt{hc/G}$ is 2.2×10^{-8} kg. It looks as though only the Planck mass scale is currently accessible to human technology, and perhaps we will not understand quantum gravity until we can access the Planck length and time scales.

Chapter 6

The twin paradox in relativity

THE twin or clock paradox is an important result of relativity which requires careful deliberation by students. In the standard resolution, the traveling twin ages less than the stay-at-home twin. In this chapter, we show that this resolution is incorrect because it rests on untenable assumptions. The new resolution also avoids logical conclusions necessitated by the kind of time travel allowed by the standard resolution.

A. The problem

The twin paradox in relativity has a history almost as long as the theory of relativity itself. It was originally proposed by Einstein as a *gedanken* experiment to highlight the fact that an observer sees a moving clock going at a slower rate than a clock at rest. Due to the fact that this is an entirely symmetric effect (in that each clock sees the other clock ticking slower), the paradox arises when two identical synchronized clocks are temporarily separated into different Lorentz frames and then brought back together. Applied to twins, the paradox starts with identical twins on Earth. One of the twins then accelerates away on a rocket, moves away from Earth at a constant velocity u for a time $T/2$, fires rockets to accelerate again so that his velocity changes to $-u$, moves at the constant velocity u towards the Earth for a time $T/2$, and decelerates to a stop on reaching the Earth again. The acceleration times are assumed to be negligible compared to T. The paradox arises because both twins observe the other to be aging slower during the period of uniform relative motion. Are they the same age when they meet again or is one of them younger, and if so, which one and by how much?

Over the years, the paradox has been discussed extensively in many books and articles. The American Journal of Physics—a journal devoted to the teaching of physics—has carried a large number of articles on the paradox, highlighting the difficulty in conveying this concept to first-time students of relativity. It still remains one of the most puzzling aspects of the theory of relativity.

B. The standard resolution

In the standard resolution, as presented in many textbooks on relativity, both twins conclude that the traveling twin who *accelerated* is younger. The argument proceeds as follows. The Earth-bound twin always remains in an inertial frame and therefore his observation that the other twin is aging slower is *correct*. On the other hand, the rocket-bound twin sees his brother age slowly during the time when the relative velocity is constant, but sees a *sudden jump* in his brother's age during the short acceleration phase when he is not in an inertial frame. Thus, the change of inertial frames results in a jump in age. The use of the word sudden, which is standard terminology in these discussions, is not meant to imply that the change is discontinuous but only to mean that the change happens during the relatively short acceleration phase.

Since the paradox involves an acceleration phase for at least one of the twins, it is useful to bring in some of the framework of general relativity to analyze this problem. A mathematical analysis along these lines is presented by R. C. Tolman in his book *Relativity Thermodynamics and Cosmology.*[*] His main argument is that the jump in age seen by the rocket-bound twin during the acceleration phase is due to the *equivalent gravitational shift* between clocks placed at different points in a uniform gravitational field. By the equivalence principle, the acceleration can be viewed equally well as arising due to the turning on of a uniform gravitation field. If the distance between the twins is D, the acceleration of the rocket is g, and the duration of acceleration (in the rocket frame) is τ_R, then in this time the accelerated twin will perceive his Earth-bound brother to age by an amount

$$\tau_E \approx \tau_R \left(1 + \frac{gD}{c^2} \right)$$

where the $\approx$ sign indicates that the relation is correct to first order. This change can be very large compared to τ_R for large values of D. In other words, the short acceleration time in the rocket frame is equal to a large time in the Earth frame.

Using this idea, we can calculate the ages of the twins from both viewpoints. For the Earth-bound twin, if the time of uniform relative motion is T_E and

---

[*]R. C. Tolman, *Relativity Thermodynamics and Cosmology* (Dover Publications Inc., New York, 1987).

the three acceleration phases have negligible duration, then his age has increased by T_E, while his brother's age T_R is shorter and related to T_E by:

$$T_E = \frac{T_R}{\sqrt{1 - u^2/c^2}} = T_R \left(1 + \frac{u^2}{2c^2} + \cdots \right) \tag{6.1}$$

From the viewpoint of the rocket-bound twin, the equivalent gravitational shifts during the three acceleration phases should also be taken into account. During the initial acceleration away from Earth and the final deceleration upon returning to Earth, the twins are almost at the same gravitational potential and the differential shift is negligible. However, during the intermediate turn-around acceleration, the twins are widely separated and the shift is large. If the travel time (in the rocket frame) before turn around is $T_R/2$, then their separation is $D = uT_R/2$. Since the velocity changes by $2u$ in a time τ_R, the acceleration is $g = 2u/\tau_R$. Therefore, the age of the Earth-bound twin as seen by his brother is:

$$T_E = T_R\sqrt{1 - u^2/c^2} + \tau_R \left(1 + \frac{T_R}{\tau_R} \frac{u^2}{c^2} \right)$$

$$= T_R \left(1 - \frac{u^2}{2c^2} + \cdots \right) + \tau_R + T_R \left(\frac{u^2}{c^2} \right) \tag{6.2}$$

Neglecting the τ_R term, this is the same relation as Eq. (6.1)—thus both twins agree that the rocket-bound twin is younger, and by the same amount (at least to leading order).

We thus see that the successful resolution of the paradox from the viewpoint of the traveling twin rests on the statement that the accelerated clock will see the faraway clock go at a slower rate due to the equivalent gravitational shift predicted by general relativity. In the words of Tolman

> *The solution thus provided for the well-known clock paradox of the special theory . . . has been made possible by the adoption of the general theory of relativity.*

The above analysis also shows why the following statement from Boughn's paper[*] is incorrect

[*]S. P. Boughn, "The case of the identically accelerated twins," *American Journal of Physics* **57**, 791–793 (1989).

> *It has often been pointed out that while the acceleration of one twin is the key to the resolution of the paradox, it is wrong to suppose that reduced aging is a direct result of acceleration. The age difference of the twins is proportional to the length of the trip while the period of acceleration is determined only by how long it takes to turn around and is independent of the length of the trip and, hence, the final age difference of the twins.*

Equation (6.2) shows clearly that the age difference is indeed proportional to the length of the trip T_R, while the period of acceleration τ_R cancels out. In other words, changing the period of acceleration changes the value of g exactly by the amount required to produce an age difference proportional to the length of the trip.

Some authors argue that there is no need to bring in the framework of general relativity to resolve the paradox. However, even when the analysis stays within the domain of special relativity, these authors conclude that there is a sudden jump in age of the far-away twin during the acceleration phase. To quote from the classic textbook *Spacetime Physics: Introduction to Special Relativity* by Taylor and Wheeler (page 130)[*]

> *This 'jump' . . . is the result of the traveler changing frames. And notice that the traveler is unique in the experience of changing frames; only the traveler suffers the terrible jolt of reversing direction of motion. In contrast, the Earth observer stays relaxed and comfortable in the same frame during the astronaut's entire trip.*

Despite the above arguments, Tolman's general relativistic analysis has two features that will prove useful to us:

1. The equivalent gravitational shift when the two clocks are nearby is negligible, so that the two observers agree on the duration of the acceleration times in the first and third phases. In other words, a local inertial observer can calculate the acceleration time in the non-inertial frame.

2. The equivalent gravitational shift when the two clocks are widely

---

[*]E. Taylor and J. A. Wheeler, *Spacetime Physics: Introduction to Special Relativity*, 2nd ed. (W. H. Freeman and Company, New York, 1992).

separated provides a way of quantitatively calculating the 'jump' due to the change in frames.

Furthermore, Tolman's analysis shows that it is not the acceleration (or the terrible jolt of changing frames) *per se* but only the acceleration which happens when the twins are separated that is important for the resolution of the paradox. In fact, this shows that a further simplification is possible—*the accelerating twin need not really complete the return journey*. It is sufficient if he just decelerates to come to rest with respect to his far-away twin, since their simultaneous ages can now be compared without ambiguity, although they will be at different locations. The initial acceleration which both twins experienced (but at the same location) is not relevant.

1. Identically accelerated twins

We next consider a well-known variant in which both twins experience a *symmetric* acceleration phase but one that still results in asymmetric aging because they are spatially separated. Consider twins who are moved to widely different locations in the Earth frame after birth. Then they get on to spaceships and accelerate identically by firing identical rockets. After the acceleration phase, they are both moving in the same direction with identical velocity of u. Thus they are at rest with respect to each other and their (simultaneous) ages can be compared without ambiguity. There is no doubting the fact that they have gone through identical experiences with respect to the acceleration phase. However, according to the earlier equivalence-principle analysis, the twin on the left would see the one on the right to be older after the acceleration phase. The only asymmetry is that the direction of acceleration is towards the other twin, so that the left twin would be older if they both accelerated to the left instead. In the language of *Spacetime Physics* (page 117)

> ... *the twins are both jumping on to a moving frame at different locations.*

The analysis using the equivalence principle now proceeds as follows. Let the duration of the uniform acceleration phase in the frame of the left twin be τ_L. Then the acceleration is $g = u/\tau_L$, so that the equivalent duration for the right twin at a distance $x_\circ$ away is

$$\tau_R = \tau_L \left(1 + \frac{ux_\circ}{\tau_L c^2}\right)$$

To give some concrete numerical values, let us take τ_L to be 1 yr and $ux_\circ/c^2$ also to be 1 yr. Thus the left twin concludes that his age after the acceleration phase has increased by 1 yr while that of his twin has increased by 2 yrs. From the point of view of the right twin, the acceleration is $g = u/\tau_R$, but now he is at the higher gravitational potential so that

$$\tau_L = \tau_R \left(1 + \frac{ux_\circ}{\tau_R c^2}\right)^{-1}$$

Hence he will reach *the same conclusion* that the twin on the left is younger. But, since the duration of the acceleration phase *in the rest frame* is the same for both twins, τ_R is 1 yr and the age of the twin on the left has increased only by 0.5 yr. In other words, while the twins will agree that the twin on the right is older, they will disagree on the amount of aging. The fact that the acceleration phase lasts for the same (proper) time seems correct because it depends on the amount of fuel and the rate of burning, which is taken to be identical for the two rockets. This is also consistent with point 1 of Tolman's argument, namely that a *locally* inertial observer can verify that the duration of an acceleration phase in a non-inertial frame is identical.

2. Incorrect use of the equivalence principle

The analysis in the previous section shows that the use of the equivalence principle in a well-known variation of the twin paradox—the case of the identically accelerated twins—results in an inconsistent answer. It therefore begs the question as to whether the equivalence principle—the bedrock of general relativity and undoubtedly correct—is being used properly in this case.

The answer is a resounding no, but before that let us consider the statement of the principle from Weinberg's book on *Relativity and Cosmology* (page 68)[*]

---

[*]S. Weinberg, *Gravitation and Cosmology: Principles and Applications of the General Theory of Relativity* (John Wiley & Sons, New York, 1972).

> *... at every space-time point in an arbitrary gravitational field it is possible to choose a "locally inertial coordinate system" such that, within a sufficiently small region of the point in question, the laws of nature take the same form as in unaccelerated Cartesian coordinate systems in the absence of gravitation.*

The key word in the above statement is *locally*, defining a region which is "sufficiently small" so that the gravitational field is constant and tidal effects can be ignored. On the other hand, the gravitational redshift is an explicitly *nonlocal* phenomenon, relating to the relative rates of two clocks placed at different potentials in a gravitational field.

In order to see explicitly that the equivalent shift cannot be used in the twin-paradox situation, we consider the origin of the shift as discussed in Tolman's book. Consider two identical clocks, placed left and right and separated by a distance D, in a frame being uniformly accelerated to the right at a rate g. Let the first clock emit a photon at time $t = 0$. Traveling at c, the photon reaches the second clock after a time D/c. In this time, the second clock has picked up an additional speed of gD/c, so that the first-order Doppler effect yields for the relative rates

$$\frac{\tau_R}{\tau_L} = 1 + \frac{gD}{c^2}$$

exactly as expected for the gravitational redshift in a uniform gravitational field. Thus, the physical basis for the shift in an accelerated frame is the normal Doppler effect. But this derivation is valid in a frame undergoing constant acceleration that lasts forever, or at least as long as it takes for a photon to reach from one clock to the other. This is explicitly not the case in the twin-paradox situation, where the duration of the acceleration phase is negligibly small and certainly not long enough for a photon to cover the distance between the twins. The increased relative velocity can be at most u (depending on the instant at which the photon was emitted), and definitely not gD/c. This shows clearly that the equivalent shift cannot be applied to the short acceleration phase of the twin-paradox kind.

The gravitational redshift in a gravitational field is a real effect. On the surface of the earth its magnitude is very small and corresponds to a relative shift of 1×10^{-16} per meter of height difference. Despite this, it was measured in a remarkable experiment by Pound and Rebka in a tall building at Harvard University.[*] The source in the experiment was a solid containing iron nuclei that emitted recoilless gamma rays, recoilless because it used the Mössbauer effect. This allowed them to measure a small shift of 3 parts in 10^{14}, corresponding to a height difference of 30 m.

---

[*]R. V. Pound and J. G. V. Rebka, "Apparent weight of photons," Phys. Rev. Lett. **4**, 337–341 (1960).

C. The correct resolution

The correct resolution to the paradox is that both twins age by the same amount, which is equal to the higher value as calculated by the Earth-bound twin who is always in an inertial frame.

This will also give the correct answer for the case of the identical accelerated twins considered in the previous section. Both the twins will age by the same amount of 1 yr, *independent of the direction of acceleration.* The same can be measured by a locally inertial observer, as required by relativity.

1. Experimental realization

An exact experimental realization of the twin-paradox effect—a shift of gD/c^2 for a given distance D between the twins—would be a repeat of the Pound-Rebka experiment, but with the source and observer at the same height and *with the absorber accelerated towards the source at a constant rate g*. Since the source and absorber are at the same gravitational potential, there will be no frequency shift due to the gravitational redshift. However, the twin-paradox effect predicts a shift of gD/c^2 because the absorber is being accelerated. Since the shift depends linearly on the distance D, one can increase the sensitivity of the experiment by increasing D. Whereas the original Pound-Rebka experiment was done with a height difference of 30 m, the distance on the ground can easily be increased by a factor of 2 or more. While the above experiment has not been done, it is unlikely to give a positive result because, if real, it would allow some kind of time travel. This raises logical questions like "*Will I be born?*" if I go back in time and kill my parents.

We also consider here an experiment[*] that is often quoted as an experimental verification of asymmetric aging in the twin paradox (for example, *Spacetime Physics* page 134). This is also an experiment by Pound and Rebka, but where the frequency of the recoilless gamma rays were studied as a function of temperature. This experiment was done primarily to check for sources of systematic error in their gravitational redshift experiment. They did find a shift in the frequency of the gamma

---

[*]R. V. Pound and J. G. V. Rebka, "Variation with temperature of the energy of recoil-free gamma rays from solids," Phys. Rev. Lett. **4**, 274–275 (1960).

ray as the temperature was increased. However, this is not a test of the twin-paradox effect but only a measurement of the second-order Doppler effect, because the increased temperature results in an increased average value of u^2. Indeed, if we could transform to the frame of the oscillating iron nuclei, the laboratory clocks would appear to go slower symmetrically. This is a manifestation of time dilation similar to the *apparently* long lifetime of particles decaying in flight as observed from a stationary frame. Nobody will argue that particles in flight age slower because of this apparent dilation.

2. Discussion

We now discuss what is the logical fallacy in the standard resolution to the twin paradox. The fallacy is to ascribe a *real* change to the *apparent* effect of time dilation between relatively moving frames. This is akin to saying that the well-known phenomenon of Lorentz contraction is real and causes a real change in the length of a moving rod. The length of a rod is determined by interatomic forces and is not dependent on its motion with respect to some arbitrary frame, while the Lorentz contraction is an apparent effect due to the relativity of simultaneity. Similarly, the age of a person (or clock) is determined by physical processes that control decay rates. Moving one person temporarily to another Lorentz frame cannot change this rate.

Indeed, it is instructive to see that the differential aging in the twin-paradox effect arises from the fact that the distance to the turn-around point appears Lorentz contracted in the rocket-bound frame and hence takes a shorter time to travel. This leads to the unlikely conclusion that travel to a distant star is possible if we travel close to the speed of light. The nearest star (outside the solar system) is 4.5 light years away, so that an astronaut traveling at $0.9c$ would take 5 yrs (in the Earth frame) to get there. But the twin-paradox effect predicts that the astronaut would have aged only by $5\sqrt{1-0.9^2} = 2.18$ yrs [Eq. (6.1)]. From the Lorentz contraction point of view, the distance to the star appears shorter by the same factor, and therefore takes a shorter time to travel. This suggests that the distance to any faraway object can be made arbitrarily small by traveling sufficiently close to the speed of light. Not at all likely.

It is also important to realize that the gravitational redshift is a real effect on the relative rates of clocks placed at different gravitational potentials. The curvature of spacetime in the presence of a gravitational field has a physical effect on natural processes so that there is a difference in clock rates when one clock is compared to the other, though no local measurements on one clock will show any change because all processes are affected the same way. Therefore, if one twin climbs up to the top of a mountain (i.e. to a higher gravitational potential) while his brother stays at the base, lives atop for one year, and then comes back down to rejoin his brother, he will find that his age is slightly higher. But this is because the curved spacetime near the Earth's surface causes a real change in the relative heart rates of the two brothers—mere acceleration in flat spacetime for a short time cannot cause such a change, as required in the twin-paradox effect.

Chapter 7

Robert Dicke and atomic physics

ROBERT DICKE, the person featured in this chapter, was a pivotal figure in 20th century physics. Apart from his contributions to atomic physics discussed here, he made fundamental contributions to several other areas of physics. In particular, he made a calculation of the temperature of the afterglow left over from the big bang signaling the birth of our universe—called the cosmic background radiation—made an estimate that the temperature was in the microwave regime, and hence amenable to experimental technology. As luck would have it, the same was accidentally discovered by Penzias and Wilson working at Bell Labs (just a few miles away from Princeton, where Dicke was a professor of physics), for which Penzias and Wilson got the Nobel Prize. Dicke was also instrumental in developing the Brans-Dicke theory of gravity, which provided an alternative to the general theory of relativity and hence made it testable. This is because a scientific theory becomes testable only if there are alternatives which predict different results, so that experiments can decide in favor of one or the other. Needless to say, general relativity has come out with flying colors after every test.

A. Lineshapes and linewidths

Spectroscopy is the name given to the process of measuring the frequency response of a system—its *spectrum*. In atomic physics, this usually means measuring one of the resonance lines of an atom, i.e. a transition from one state to another. It is well known that no resonance line is infinitely narrow— even in theory—because spontaneous emission introduces a finite lifetime to the excited state, and therefore (by the time-frequency uncertainty principle) a natural width to the transition. In practice, a number of other processes, such as collisions, Doppler effect, field inhomogeneities, light intensity (power broadening), etc., act to increase the width of resonance lines. It is important to understand how these effects manifest themselves in the shape of the resonance curve (the "lineshape") so that one can do a curve fit to the observed spectrum and thereby find the resonance frequency ω_0.

Line broadening mechanisms are broadly classified into two— *homogeneous* and *inhomogeneous*—depending on whether the mechanism affects the lineshape in a similar way for each particle (homogeneous), or whether the broadening arises from a random shift in the resonance frequency for different particles causing a widening for an ensemble of particles (inhomogeneous). Thus, the natural linewidth and collision-induced effects are examples of homogeneous broadening, while the Doppler effect and field inhomogeneities result in inhomogeneous broadening.

Mathematically, the lineshape is quite different for the two broadening mechanisms. Homogeneous broadening arises from random changes in the phase (or coherence) of the radiation from the particles, and therefore results in a *Lorentzian* lineshape. The normalized Lorentzian function is given by

$$L(\omega) = \frac{1}{\pi} \frac{\Gamma/2}{(\omega - \omega_0)^2 + (\Gamma/2)^2}$$

where Γ is called the full width at half maximum (FWHM). For the natural linewidth, Γ is the inverse of the excited-state lifetime. Since Γ is the linewidth in angular frequency units, it has to be divided by 2π to get the linewidth in normal frequency units.

Inhomogeneous broadening results from the random perturbations in the resonance frequency of the particle which follows a Gaussian (or normal) distribution, and thus produces a *Gaussian* lineshape. The normalized Gaussian function is given by

$$G(\omega) = \frac{1}{\sqrt{2\pi\sigma^2}} \exp\left[-\frac{(\omega - \omega_0)^2}{2\sigma^2}\right]$$

where σ is the width of the Gaussian (again in angular frequency units).

1. Lineshapes of confined particles

Confined or trapped particles offer the ultimate in spectroscopic resolution because

1. The trapped particle can be cooled with lasers or cryogenically to a temperature of order 1 mK, so that the second-order Doppler effect is below 10^{-16}.

2. With proper vacuum, the effect of collisions can be virtually eliminated.

3. The first-order Doppler effect can also be suppressed because the spectrum of a trapped particle consists of an unshifted central line and sidebands separated by the oscillation frequency in the trap, and it is possible to just address the recoilless central line, as in the case of the Mössbauer effect or in the use of buffer gases.

To see this last point more clearly, consider a particle trapped in a harmonic potential. If the particle is oscillating in the trap at a frequency ω_t and amplitude x_0, then its instantaneous phase at a faraway detector is

$$\phi(t) = -kx_0 \sin \omega_t t - \omega_0 t$$

where $k = 2\pi/\lambda$ is the wavevector. The instantaneous frequency of the emitted wave is therefore

$$\omega = -\dot{\phi}(t) = kx_0\omega_t \cos \omega_t t + \omega_0$$

The first term can be recognized to be the usual Doppler shift of kv. In the language of electrical engineering, the above equations show that the wave is undergoing *frequency modulation* (FM)* with a modulation index

$$\beta = kx_\circ$$

Any standard textbook will tell you that the amplitude of a wave undergoing FM modulation with an index β is

$$a(t) = \sum_{n=-\infty}^{\infty} (-1)^n J_n(\beta) \cos[(\omega_\circ - n\omega_t)t]$$

where J_n's are n^{th}-order Bessel functions. Thus the spectrum consists of a central line at $\omega_\circ$, and an infinite set of sidebands spaced uniformly at multiples of the oscillation frequency ω_t but with progressively smaller amplitude. The spectral intensity of the emitted light is proportional to

$$I(\omega) \propto \sum_{n=-\infty}^{\infty} J_n^2(\beta)\delta(\omega - \omega_\circ - n\omega_t)$$

The effect of β on the spectral intensity is seen graphically in Fig. 7.1. The height of the central peak and three sidebands are shown in the figure for two values of β. For $\beta = 0.5$, only the $n = \pm 1$ sidebands are significant, while the others are practically zero. For $\beta = 0.1$, even these sidebands are almost zero.

---

*The other kind of modulation used in radio communication is *amplitude modulation* (AM), where the amplitude of a carrier wave is modulated by the signal that needs to be transmitted.

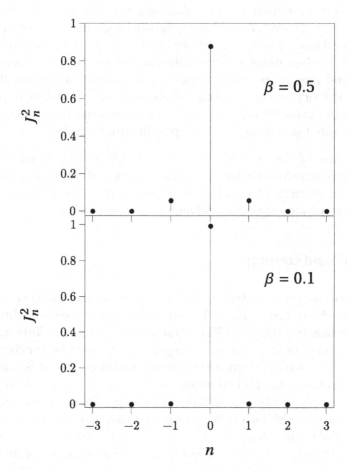

Figure 7.1: Spectral intensity of central peak and sidebands in FM modulation for two values of the modulation index β. The sidebands are practically zero for $\beta = 0.1$.

2. Tight confinement—the Lamb-Dicke regime

The discussion above shows that the most dramatic effects of confinement will be seen when the particle is confined to a size that is smaller than the wavelength of the emitted light, called tight confinement. In other words, $x_o \ll \lambda$ or $\beta = kx_o \ll 1$. From Fig. 7.1, the sidebands are almost completely suppressed when $\beta \ll 1$, and only the central unshifted line survives. This is called *recoilless emission* because the particle has the same momentum before and after the emission, and the momentum of the emitted photon is provided (or taken up in the case of absorption) by the whole trap. This is analogous to the Mössbauer effect in solids where the recoil momentum of the emitted gamma ray is taken up by the whole crystal.

The regime of tight confinement is particularly easy to achieve for microwave transitions in ion traps, where the wavelength is much larger than the trap size. Such confinement plays an important role in precision spectroscopy, atomic clocks, and quantum computation.

3. Sideband cooling

Trapped ions can also be laser cooled using what is called *sideband cooling*. The basic idea is seen in Fig. 7.2. The external degree of freedom of the ion in the harmonic trap (i.e. its oscillatory motion) is quantized to form discrete energy levels. The levels are equally spaced for a harmonic oscillator. On top of this, the internal degrees of freedom form the energy levels of the ion. Thus, the ground and excited states form two sets of harmonic-oscillator levels spaced apart by ω_{ge}, which is the energy difference between the two states. For sideband cooling, the laser is tuned to the lower motional sideband of the transition, i.e. $\omega_c = \omega_{ge} - \omega_t$. After absorbing such a laser photon, the ion will emit by spontaneous emission most probably at ω_{ge}. Thus the ion loses one quantum of trap motion in each absorption-emission cycle. This cooling works best when the natural linewidth of the upper state is much less than ω_t, so that the uncertainty in the levels of the upper state is much smaller than their spacing. Under these conditions, the ion can be cooled to the lowest quantum state of the trap and one can realize tight confinement.

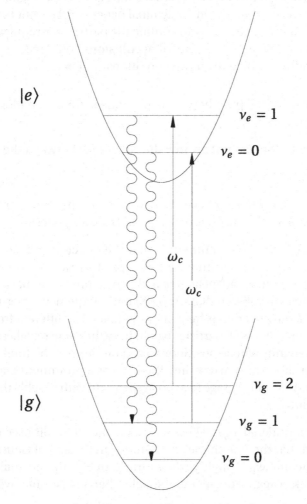

Figure 7.2: Laser cooling of harmonically-trapped ions by tuning the laser to the lower motional sideband. The ground and excited state of the ion are quantized into equally spaced motional levels. The ion loses one quantum of motion in each absorption-emission cycle.

4. Dicke narrowing

The regime of tight confinement is called the Lamb-Dicke regime in honor of Dicke's contribution. But in his seminal paper,[*] Dicke went beyond and argued that it is not necessary to confine the particle harmonically to see significant narrowing. He showed that **collisions** could reduce the usual Doppler width significantly if two conditions were met:

1. The mean free path between collisions is much smaller than the wavelength λ.

2. The collisions do not destroy the coherence between the radiating states.

Under these conditions, Dicke showed that the spectrum consists of a narrow "recoilless" peak on top of a broad Doppler pedestal.

The essence of Dicke's argument is that the effect of collisions can be viewed as a series of one-dimensional traps of size a, where $a \ll \lambda$ since the scale of a is set by the mean free path. The atom moves back and forth between the two walls, so that the oscillation frequency varies with a. In other words, the radiation is frequency modulated at different frequencies depending on a. Thus all particles would have the same recoilless line and a set of sidebands whose positions varied randomly. The total emission would then show this narrow unshifted line as a prominent feature, and the sidebands would average to a broad spectrum with roughly the original Doppler width.

Dicke also considered a gas where $a \ll \lambda$ and the atoms diffused randomly. He showed that the line shape was Lorentzian instead of Gaussian. The width of the line was roughly $2.8\, a/\lambda$ times that of the normal Doppler width, which is again considerably smaller than the Doppler width.

The only thing that Dicke did not say in this paper was that the recoilless peak could be seen more easily for a nuclear transition from nuclei localized in a solid. Otherwise, the Mössbauer effect would have been discovered several years earlier and would have been called the Dicke effect, and Dicke would have won a Nobel prize for the work.

---

[*]R. H. Dicke, "The effect of collisions upon the Doppler width of spectral lines," Phys. Rev. **89**, 472–473 (1953).

B. Coherence

Coherence is a term that is used to describe a process where amplitudes add with a relative phase and one observes a physical quantity that is proportional to the square of the total amplitude. A familiar example is the interference pattern from two slits one sees in classical optics. The light coming from the two slits is coherent, so that at any point on the screen one adds the E field amplitudes, while the observed light is proportional to the square of the total E field at that point. In quantum mechanics, coherence and interference arise when the wave amplitudes have to be added.

In atomic physics, the term coherence is used for two distinctly different phenomena: (i) those which occur in a *single* atom when two states have a relative phase between them, and (ii) those which occur when there is more than one atom radiating so that radiative coupling between the atoms becomes important. The first kind of coherence leads to phenomena such as quantum beats (in which the radiation rate oscillates in time) and level crossing (in which interference between two nearly degenerate states modifies the fluorescence intensity and polarization). The second kind dominates when the atoms are localized to a spatial region that is smaller than a wavelength of the radiation. Then, radiative coupling between the atoms becomes important and the phase of the field between the radiators does not vary much, i.e. retardation effects can be neglected.

Dicke discusses this second kind of coherence, namely the collective behavior of N particles decaying by spontaneous emission, in a landmark paper.[*] If the radiation rate of a single particle is I_1, one expects that the rate for the entire system would be just $N I_1$. But this is incorrect because it ignores the fact that all the particles are interacting with a common radiation field.

1. Coherence between two spin-1/2 particles

To illustrate this point, Dicke considers the coupling between two two-level systems. The canonical example of a two-level system is a spin-1/2 particle

---

[*]R. H. Dicke, "Coherence in spontaneous radiation processes," Phys. Rev. **93**, 99 (1954).

(e.g. neutron) in a magnetic field.* If a neutron is placed in a uniform magnetic field in the higher of the two spin states, then in due course of time it would decay to the lower state by spontaneously emitting a photon via a magnetic dipole transition. Thus the probability of finding it in the upper state would decay exponentially to zero.

Now consider what happens if a second neutron in the *ground state* is placed close to the first excited one. Here, close means at a distance small compared to the radiation wavelength, but large compared to the interparticle dipole-dipole interaction length. If the independent radiator hypothesis is correct, the radiation from the first particle would not change. But, in reality, the radiation would be strongly affected because the transition probability would fall exponentially to one-half and not zero.

To see this, consider the combined states of two spin-1/2 particles. Their individual spin-angular momenta add to form what are called singlet and triplet states. The singlet state has total angular momentum of zero, and therefore a single magnetic sublevel. On the other hand, the triplet state has total angular momentum of 1, and hence three sublevels. Remember that the total angular momentum characterizes the eigenstate of the square of the spin operator S^2, while the sublevels are eigenstates of the z component of the spin operator S_z. If we denote their simultaneous eigenstate by $|s, m\rangle$, then

$$S^2 |s, m\rangle = s(s + 1) |s, m\rangle$$
$$S_z |s, m\rangle = m |s, m\rangle$$

One can think of the singlet state as one where the two neutrons are anti-aligned, and the triplet state as one where the two neutrons are aligned.

Therefore, when the second unexcited neutron is added, the initial state (with $m_s = 0$) is an equal superposition of the singlet and triplet states, namely the $|0, 0\rangle$ and $|1, 0\rangle$ states. The singlet and triplet sets do not couple to each other, thus only the triplet state can decay to the lower energy state while the singlet state is stable. Hence, after a long time, there is still a probability of one-half that no photon has been emitted. Indeed, if there is no photon emitted after a long time, we can be sure that the neutrons are

---

*In fact, there is a theorem, called the Feynman-Vernon-Hellwarth theorem, which states that the dynamics of any two-level system can be mapped to that of a spin-1/2 particle in a magnetic field.

in a singlet state but cannot say which neutron is in the excited state. One can also show that the radiation rate is double that for just a lone excited neutron.

2. Dicke superradiance

From the above discussion, it is clear that coherence between two radiators can strongly influence the spontaneous radiation rate. Dicke now extends this discussion to the coherent radiative behavior of N two-level systems. Unlike in his paper, we will consider these systems to be spin-1/2 particles so that what he calls the cooperation operator $\mathbf{R}$ becomes the familiar angular momentum operator. We have,

$$\mathbf{R} = \sum_{i=1}^{N} \mathbf{S}_i$$

as the total angular momentum operator. The eigenvalue of $\mathbf{R}^2$ is

$$r(r+1) \leq \frac{N}{2}\left(\frac{N}{2}+1\right)$$

The eigenvalue of the z component of $\mathbf{R}$ is

$$m = \sum_{i=1}^{N} m_i = \frac{1}{2}(n_+ - n_-)$$

where n_+ and n_- are the numbers of up and down spins, respectively. Thus, r represents how much the spins are aligned while m represents the projection of the total spin along the z direction.

Using this formalism, Dicke shows that the matrix element governing the interaction with the field (i.e. the raising operator to go from $m-1$ to m) is proportional to $[(r+m)(r-m+1)]^{1/2}$. Hence, the radiated intensity is

$$I = I_1(r+m)(r-m+1)$$

where I_1 is the rate for a single system. This is much larger than NI_1 when r is big (which can be as large as $N/2$ if all the spins line up) and m is small (which is 0 if the total spin precesses in the xy plane). Such states with enhanced radiation rates are called *superradiant* states. For example, for the case when $r = N/2$ and $m = 0$, $I \approx N^2 I_1$, which is much larger than $N I_1$ and is in accord with our earlier discussion on the effect of coherence for N particles. On the other hand, if r is small (can be $1/2$ when N is odd and 0 when N is even), then $I = I_1$ or 0 depending on whether N is odd or even. These states are *subradiant* because they radiate less rapidly compared to N independent systems. The intermediate case when both r and m are big, say $N/2$ and r, the rate is $I = N I_1$, the same as for a set of N independent radiators.

Dicke suggests two ways of producing a superradiant state with $r \approx N/2$ and $m \approx 0$.

1. Cool the system so that it is in its ground state of $r = N/2$ and $m = -r$. Then give it a pulse to bring it to the xy plane where $m = 0$. Such a pulse is called a $\pi/2$ pulse in NMR since it changes the angle of the magnetic moment by $\pi/2$.

2. Put the system in its extreme excited state with all the particles in the upper level, i.e. with $r = N/2$ and $m = r$, and then just *wait*. At first, the system will decay as independent radiators $I = N I_1$. Since the selection rule for the transitions is $\Delta r = 0$ and $\Delta m = -1$, the system will radiate faster and faster until $m \approx 0$. At this point, the intensity will be $N/2$ times larger, which can be quite dramatic if N is large.

In his paper, Dicke also points out that the rate of stimulated emission/absorption is not enhanced by N^2, but only by N, even for the state with $r = N/2$. Thus, his analysis demonstrates a surprising result—the rate of *spontaneous emission*, which is a self-induced and random process for a single atom, shows coherence effects for a suitably localized ensemble of atoms.

Chapter 8

The 1997 Nobel Prize in Physics—Laser Cooling and Trapping

THE 1997 Nobel Prize in Physics was awarded to Steven Chu of Stanford University (USA); Claude Cohen-Tannoudji of College de France (France); and William Phillips of National Institute of Standards and Technology, Gaithersburg (USA). They have been cited for their seminal work in *the use of laser light to cool and trap atoms*. This award represents the recognition of more than two decades of work by many leading atomic physicists around the world. Indeed, several of the key ideas in this field have not originated in the laboratories of the above scientists. However, they have been chosen for the award because, in a remarkably productive period in the late 1980's, these three researchers, working independently, discovered new experimental results that surpassed the predictions of existing theories of laser cooling. Soon after, they also developed new theories that explained the strange results and helped zero in on the best conditions for laser cooling. It is one of those rare instances in physics where experiments have worked better than anticipated, and is a tribute to careful experimentation.

A. Laser cooling

The term "laser cooling" sounds like an oxymoron since lasers are normally associated with intense heat—the image of a red beam cutting through the thick metal door leading to the vault of a bank, for example. Despite this, lasers can be used to cool a cloud of atoms to extremely low temperatures, and it happens because in the quantum world of atoms, the laws of physics are very different. To understand this better, let us review some of the relevant properties of light and its interaction with atoms.

As we all know from basic electrodynamics, light (at a classical level) is a manifestation of the wavelike solution of Maxwell's equations. We are familiar with many of its properties that result from this wave nature—reflection, refraction, diffraction, interference, etc. There is however another property of waves that is not commonly associated with light, and that is the *Doppler effect*. This is better known in the case of sound waves and is the familiar change in the pitch (or frequency v) of a source when there is a relative motion between the source and observer. The fractional change in frequency is proportional to the ratio of the source velocity v to the speed c of waves in the medium, i.e.

$$\frac{\Delta v}{v} = \frac{v}{c}$$

Since the speed of sound in air is 330 m/s (1200 km/hr) and the speed of light is 3×10^8 m/s (1 million km/hr), we can see why we are more familiar with the Doppler effect for sound waves. Typical terrestrial speeds on human scales can reach the speed of sound and even break the sound barrier, but the speed of light is a long way off. For example, a train travelling at 100 km/hr is moving at 10% the speed of sound, but a negligible fraction of the speed of light. Therefore, while its whistle pitch might change as it zooms past us, the red light in the guard van always looks red!

However, with atoms in a gas the situation is different. At room temperature, a typical gas atom is shooting around at 500 m/s.* Furthermore, gas atoms detect the incident light frequency with much greater sensitivity than our

---

*You could have actually guessed this without any calculation. Remember that sound waves in air are pressure fluctuations that are transmitted from one point to another by the motion of gas molecules. Therefore, their mean speed should be of the same order as the speed of sound, or about 330 m/s.

eye due to the phenomenon of *resonance* (discussed in detail below), and therefore are more sensitive to the Doppler shift.

The next property of light is also well known since the days of the introduction of the light quantum hypothesis, namely that light waves come in packets called *photons*. We have all learnt in high school that each photon carries an energy of $h\nu$, where h is the Planck constant, and a parameter whose presence is a sure sign that quantum physics is involved. But what we are not taught in school is that each photon also carries one unit of *momentum* equal to h/λ, where λ is the wavelength of the light. This implies that each time an atom absorbs or emits a photon, by the principle of *conservation of momentum*, it experiences a momentum change equal to h/λ. Absorption causes an increase in momentum by this amount, while emission causes the same decrease in momentum.

The final property we must understand is one that governs atom-photon interactions, namely the phenomenon of *resonance*. Resonance is the reason soldiers are asked to march out of step when crossing a suspension bridge, because if their marching frequency matches the oscillation frequency of the bridge, it can cause the bridge to start oscillating at large amplitudes and collapse. The condition that there is significant energy transfer only when the drive frequency matches the internal frequency is true for atoms also. The electric field of the incident light wave jiggles the electrons in an atom, but the process is efficient at transferring energy (or photons) to the atom only when the light frequency matches the internal frequency of the atom. This result is shown in Fig. 8.1, where we plot the photon absorption rate as a function of the light frequency. The shape of this curve is a "Lorentzian", and is a universal feature of resonance phenomena. The width of the curve is the linewidth Γ of the excited state, and has a value of 10 MHz for a typical atom such as sodium.

One consequence of the Doppler shift discussed earlier is that the horizontal axis of the resonance curve in Fig. 8.1 can also be regarded as a velocity axis. Thus, if the laser is tuned to the resonance frequency of the atom in the laboratory frame, only atoms at zero velocity will absorb photons at the maximum rate given by the peak of the curve. Any atom that has a nonzero velocity in the laboratory frame will see a Doppler shifted frequency and consequently will scatter fewer photons. The narrow linewidth of the resonance curve also explains why the atoms are so sensitive to the Doppler shift. For example, a sodium atom moving with

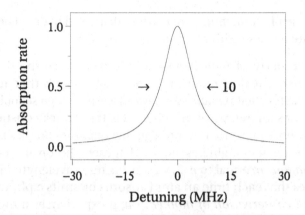

Figure 8.1: Photon scattering rate shown as a function of detuning from resonance. The lineshape is Lorentzian, which is typical for any resonance phenomenon, and the linewidth is 10 MHz.

a velocity of 30 m/s (10% of the mean velocity at room temperature) has a Doppler shift of 10 MHz, and will hardly interact with a laser beam.

It is also useful to consider the scattering process in some detail. Each absorbed photon takes the atom from the ground (or stable) state to an excited state, with a concomitant momentum kick of h/λ in the direction of the laser beam, as explained earlier. Before the atom can absorb another photon, it has to de-excite back to the ground state; this it does through a process known as spontaneous emission. The spontaneous emission is a random process and the photon can be emitted in any direction. *On the average*, this process therefore does not result in any momentum transfer to the atom. Thus, we see that each absorption-spontaneous emission cycle results in a *net* momentum transfer of h/λ to the atom in the direction of the exciting laser beam.

The Doppler cooling scheme can be understood qualitatively based on the above concepts. For simplicity, we consider a one-dimensional case where the atoms are confined to move along the x axis. The cooling configuration then consists of two identical laser beams propagating in the $+x$ and $-x$ directions, and whose frequencies are chosen to be *slightly below the resonance frequency* of the atoms. To understand the cooling process, consider an atom moving in the $+x$ direction with a small velocity. The atom Doppler shifts the $-x$ beam to a higher frequency (closer to resonance) and the $+x$ beam to a lower frequency (further away from resonance). Therefore, it preferentially scatters photons from the right beam (from Fig. 8.1) and gets more momentum kicks in the $-x$ direction which slows it down. The opposite happens for an atom moving in the $-x$ direction—it Doppler shifts the $+x$ beam closer to resonance and gets more momentum kicks in the $+x$ direction.

Thus, we have achieved a kind of frictional force that always opposes the motion. Just as friction between the ground and a rolling marble always opposes the motion of the marble and finally brings it to rest, the atom is brought to rest by the friction force from the two laser beams. A useful picture to have in mind is to imagine that you are the atom and that the laser beams are two people throwing ping-pong balls at you from both sides. If you move in one direction, say to the right, the person on the right increases his rate of throwing balls. Similarly, if you are moving towards the left, the person on the left increases his rate. The effect of the impact from many balls—each of which is very small—is to slow you down until you finally come to rest, at which point the impact rate from both sides is equal and opposite.

A more quantitative picture of this argument is obtained by considering the absorption rates from the two beams as a function of velocity. Note that the force is just the product of the absorption rate times the momentum transferred per absorption-emission cycle (this is just a restatement of Newton's second law—force is the rate of change of momentum). The result is shown in Fig. 8.2. The force from the $+x$ travelling beam is positive and centered at a negative velocity. The curve is a rescaled version of the resonance curve in Fig. 8.1, and since the frequency is below resonance, the maximum scattering rate occurs when the atom is moving in the $-x$ direction with a small velocity. Similarly, the force from the $-x$ travelling laser beam is negative (upside down version of the resonance curve) and is centered at a positive velocity. The sum of these two forces results in a net force that is linear in velocity near the origin (is proportional to velocity) and has a negative slope (always opposed to the velocity). Such a force is by definition a frictional force.

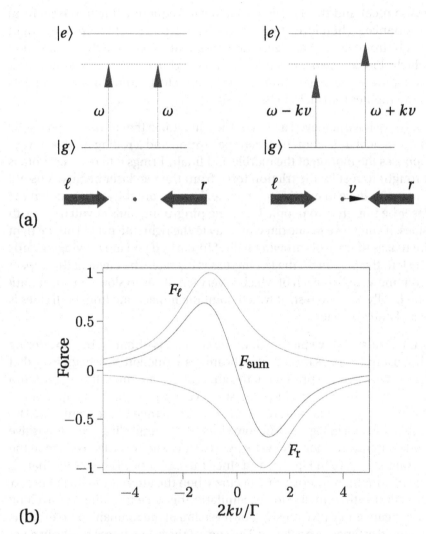

(a)

(b)

Figure 8.2: Laser cooling of atoms using optical molasses, shown in 1D for simplicity. The two laser beams are equally detuned below resonance. (a) The two forces are equal when the atom is stationary; but become unequal when the atom is moving with a velocity v, because one beam is Doppler shifted closer to resonance. (b) Forces due to the left beam, right beam, and their sum, as a function of velocity, for the configuration shown in (a). The detuning is chosen to be $\Delta = -2kv/\Gamma$, so as to achieve maximum cooling.

At this stage, two questions arise that need to be answered. Firstly, how restrictive is our analysis of the one dimensional case when extended to the three dimensions of real space? Well, it turns out that the scheme can be readily extended by having three orthogonal sets of counter-propagating beams, one each along the x, y, and the z axes. Since the velocity of the atom can be resolved into components along these three axes, each component is brought to zero by the above argument until the atom comes to rest in all three dimensions. Such a configuration of laser beams has been called *optical molasses*, to convey the idea that the atoms are moving through a viscous fluid. Experimentally, it was first demonstrated by Steven Chu and his collaborators at AT&T Bell Laboratories in 1985.

The second question that needs to answered is, "does the atom really come to rest?" Actually, this is not true for several reasons, not the least of which is that it would violate the Heisenberg uncertainty principle. What happens is that the mean velocity of the atoms is zero, but the mean squared is not. In other words, the atom is equally likely to have a speed in any direction, so that their average is zero, but the speed itself is not zero. This is a well-known result from the kinetic theory of gases. A gas at a finite temperature will have a finite root mean square velocity which, in fact, is a measure of the temperature. In our case, atoms at rest would correspond to absolute zero temperature but, in reality, they are cooled only to a finite final temperature.

The key to understanding this lies in the spontaneous emission process. We argued earlier that spontaneous emission results in no *net* transfer of momentum to the atom since it was a random event. However, each individual event still gives a momentum kick of h/λ to the atom, it is just that they average to zero. Now consider an atom that starts out at rest in our one-dimensional molasses. Each spontaneous photon gives it a random kick to the left or right. This results in a random walk in momentum space in steps of h/λ and the mean square displacement from the origin increases linearly with time. Thus the atom gets "heated" at a steady rate. Under equilibrium conditions, the heating and cooling rates are equal and a final steady state temperature is reached. A little bit of algebra shows that the lowest temperature is attained when the lasers are *detuned* half a linewidth ($\Gamma/2$) below resonance, and the lowest temperature is given by

$$T_{\min} = \frac{h\Gamma}{2\pi k_B}$$

where k_B is the Boltzmann constant. For a typical atom such as sodium, which was used for many of the experiments, $\Gamma = 10$ MHz, and the lowest temperature is 240 µK.

One important note about the spontaneous emission process. Since it is a random process, it increases the entropy (or disorder) of the photon field. Cooling the gas of atoms involves a decrease in entropy of the atoms. From the second law of thermodynamics, we know that the total entropy in any process can only increase, therefore we could not have cooled the atoms without increasing the entropy somewhere. The "somewhere" in this case is the electromagnetic field. Thus, the entropy is dissipated by the spontaneous photons which therefore form an integral part of the cooling process. Now, another well-known theorem in statistical physics, called the *fluctuation-dissipation theorem*, asserts that any process that involves dissipation will finally result in fluctuations. Applied to our problem, the laser field initially causes the momentum to dissipate but ultimately results in momentum fluctuations or heating. Thus there is no way of avoiding the heating process.

By the year 1986, researchers in many laboratories were able to demonstrate Doppler cooling using optical molasses in alkali atoms such as sodium. The energy levels of sodium are in the visible—yellow color that you see from sodium vapor lamps in the streets—and easily accessible with tunable dye lasers. The early results seemed to confirm the predictions of the above model, albeit without great accuracy. Soon after that, however, scientists started doing careful experiments to measure the velocity distribution and temperature of the cloud of atoms under varying conditions of laser detuning, laser intensity, ambient magnetic field and so on. This is when startling new results were discovered. For example, in sodium, the lowest temperature was found to be much less than the theoretically predicted minimum of 240 µK, and was found at a detuning of 20–30 MHz, not at 5 MHz ($\Gamma/2$) as predicted. The temperature was also found to be extremely sensitive to stray magnetic fields, fields which were so small that they should not have caused any perturbation in the above model.

A flurry of experimental activity followed with lower and lower temperatures being reported. The researchers accumulated an enormous amount of experimental data before sufficient clues were obtained to explain the anomalous results. They discovered that the basic flaw in the old theory was that the atom was treated too simply in assuming that it had just one ground and one excited state. In reality, each state is made up of several sublevels whose energies are shifted by varying amounts depending on the local magnetic field, light intensity, laser polarization, etc. When these factors were put in, there emerged a new theory that not only explained all the data but also predicted that, under the right conditions, temperatures of only a few µK could be achieved. The temperature limit corresponded to the recoil velocity imparted to the atom by the emission of a single spontaneous photon—the so-called *recoil limit*. This limit would *prima facie* appear to be a fundamental limit since the uncertainty in the velocity of the atom should at least correspond to the velocity imparted by the last spontaneously emitted photon. Well, after the development of these new theories of laser cooling by Chu and Cohen-Tannoudji, temperatures close to the limit were indeed measured. But the scientists have not been satisfied with that either. Since then, they have demonstrated ingenious schemes for reaching sub-recoil temperatures, and now it is believed that there is no theoretical limit to the lowest attainable temperature.

B. Magneto-optic trap—MOT

Cooling of atoms using lasers localizes atoms in momentum space—reduces their velocity spread. But this can be achieved at any point in space.[*] In order to localize them in real space—trap them—we need to provide a restoring force that pushes the atoms towards a particular point in space. This is most easily achieved by adding a quadrupole magnetic field to the 3D molasses configuration. Such a field has a linear variation along any of the three axes, and is produced using a pair of coils in anti-Helmholtz configuration. The laser beams are chosen to have opposite circular polarizations along the three axes, so that selection rules for transitions among the magnetic sublevels can be used to achieve the required differential scattering rate.

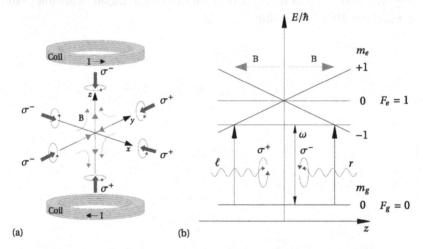

(a)   (b)

Figure 8.3: The magneto-optic trap (MOT). (a) The 3D MOT requires the addition of a quadrupole B field (produced using a pair of anti-Helmholtz coils) to a molasses configuration with opposite circular polarizations along the three axes. (b) Principle of operation of the MOT using the z axis as an example. The Zeeman shifts of the m sublevels in the linear B field, and selection rules for σ polarization, mean that the ℓ beam is closer to resonance on the left hand side, while the r beam is closer to resonance on the right hand side.

---

[*]In 3D molasses, the overlap of the three laser beams defines a localized region in space where atoms get cooled in all dimensions. However, this is not sufficient to trap them.

The MOT schematic is shown in Fig. 8.3. It is called a magneto-optic trap because it uses a combination of magnetic fields and laser light to trap atoms. The principle of the MOT can be understood by considering what happens in one dimension, along the z-axis for example. The idea is shown in part (b) of the figure. For simplicity of understanding, we consider a transition which has $F = 0$ in the ground state and $F = 1$ in the excited state.* The linearly varying B field means that the magnetic sublevels have linearly varying energy shifts, and intersect at the origin where the field is 0. The laser beam on the left has σ^+ polarization and drives transitions with selection rule $\Delta m = +1$, while the laser beam on the right has σ^- polarization and drives transitions with selection rule $\Delta m = -1$. If the laser frequency is tuned below resonance (called red detuning), then an atom on the right hand side will see the right beam closer to resonance compared to the left beam. This imbalance in scattering rates results in a net force that pushes the atoms towards the origin. The opposite happens for an atom on the left hand side—the left beam is closer to resonance compared to the right beam and again pushes the atom towards the origin. We have therefore created a restoring force that pushes atoms towards the origin—the center of the coils.

1. Applications

The MOT is the starting point for all kinds of experiments using cold atoms—both in fundamental physics and technological applications.

1. Cold atoms are perfect labs for trying to understand the laws of physics, because they move with a very small velocity and stay in the apparatus for a long time.

2. Laser-cooled atoms have been used to observe a phenomenon called Bose-Einstein condensation, which we will see in the next chapter.

3. Cold atoms have already being used to make the next generation of atomic clocks. Such clocks are important for applications ranging from satellite communication to automatic navigation by cars.

---

*This assumption is not so restrictive—numerical analysis shows that the scheme works equally well for other values of angular momenta.

4. Cold atoms could impact the emerging field of nanolithography, using a form of atom lithography where atoms are manipulated by light.

Thus the potential list of applications of cold atoms seems endless.

Chapter 9

The 2001 Nobel Prize in Physics—Bose-Einstein condensation

THE 2001 Nobel Prize in Physics was awarded to Eric Cornell of the National Institute of Standards and Technology, Boulder (USA); Wolfgang Ketterle of the Massachusetts Institute of Technology, Cambridge (USA); and Carl Wieman of the University of Colorado, Boulder (USA). They have been cited *for the achievement of Bose-Einstein condensation in dilute gases of alkali atoms, and for early fundamental studies of the properties of the condensates.* The scientists have been recognized for their pioneering work in a field that has grown explosively around the world in the past few years. Though the phenomenon of Bose-Einstein condensation (BEC) was predicted by Einstein in 1925 (based on the new statistics of Bose), it was observed only in 1995. In this chapter, we review the basic physics behind the phenomenon, the experimental techniques involved in achieving it, and highlight some of the potential applications of condensates.

A. Bose-Einstein condensation

The story of BEC begins in 1924 when the young Indian physicist S. N. Bose gave a new derivation of the Planck radiation law. He was able to derive the law by reducing the problem to one of counting or statistics: how to assign particles (photons) to cells of energy $h\nu$ while keeping the total energy constant. Einstein realized the importance of the derivation for developing a quantum theory of statistical mechanics. He argued that if the photon gas obeyed the statistics of Bose, so should material particles in an ideal gas. Carrying this analogy further, he showed that the quantum gas would undergo a phase transition at a sufficiently low temperature when a large fraction of the atoms would condense into the lowest energy state. This is a phase transition in the sense of a sudden change in the state of the system, just like steam (gaseous state) changes abruptly to water (liquid state) when cooled below 100°C. But it is a strange state because it does not depend on the interactions of the particles in the system, only on the fact that they obey a kind of quantum statistics.

In modern physics, the phenomenon is understood to arise from the fact that particles obeying Bose-Einstein statistics (called bosons) "prefer" to be in the same state. This is unlike particles that obey Fermi-Dirac statistics (fermions), and therefore the Pauli exclusion principle, which states that no two of them can be in the same state. In some sense bosons try to "imitate" each other and aggregate in a group where they can lose their identity and be all alike! With this property of bosons in mind, imagine a gas of bosons at some finite temperature. The particles distribute the total energy amongst themselves and occupy different energy states. As the temperature is lowered, the desire of the particles to be in the same state starts to dominate, until a point is reached when a large fraction of the particles occupies the lowest energy state. If any particle from this state gains some energy and leaves the group, his friends quickly pull him back to maintain their number! This is a Bose-Einstein condensate, with the condensed particles behaving like a single quantum entity.

The point at which "the desire for the particles to be in the same state starts to dominate" can be made more precise by considering the quantum or wave nature of the particles in greater detail. From the de Broglie relation, each particle has a wavelength λ_{dB} given by h/mv, where m is the mass and v is the velocity. As the temperature is lowered, the mean velocity of the particles decreases and the de Broglie wavelength increases. BEC occurs when λ_{dB} becomes comparable to the average interparticle separation. At this point, the wavefunctions of the particles overlap and they become aware of their likeness for each other! The average interparticle separation for a gas with number density n is $n^{-1/3}$, and, from kinetic theory, the mean de Broglie wavelength of gas particles at a temperature T is $h/(2\pi mkT)^{1/2}$. For the wavefunctions to overlap, this product should be of order 1. A more rigorous analysis shows that BEC occurs when the dimensionless phase-space density $n\lambda_{dB}^3$ exceeds 2.612.

In the early days, it was believed that BEC was only a theoretical prediction and was not applicable to real gases. However, the observation of superfluidity in liquid He in 1938 made people realize that this was a manifestation of BEC, even though it occurred not in an ideal gas but in a liquid with fairly strong interactions. BEC in a non-interacting gas was now considered a real possibility. The first serious experimental quest started in the early 1980s using spin-polarized atomic hydrogen. There were two features of hydrogen that were attractive — (i) it was a model system in which calculations could be made from first principles, and (ii) it remained a gas down to absolute zero temperature without forming a liquid or solid. Spin-polarized hydrogen could also be trapped using suitable magnetic fields. Each H atom behaves like a little magnet and, if it were aligned anti-parallel to the external field, it would be trapped near the point where the field is a minimum. Using cryogenic techniques, the gas was cooled to about 1 K and then loaded into a magnetic trap.

One of the major developments to come out of these efforts was the proposal in 1986 by Harald Hess, then a post-doc with Dan Kleppner at MIT, to use evaporative cooling to lower the temperature and reach BEC. The idea in evaporative cooling is to selectively remove the hottest atoms from the trap, and then allow the remaining atoms to thermalize. Since the remaining atoms have lower energy, they thermalize to a lower temperature. This is similar to cooling coffee in a cup: the hottest particles near the top evaporate and take away the heat, while the remaining particles get colder. The MIT group of Kleppner and Greytak demonstrated evaporative cooling of spin-polarized H by lowering the height of the magnetic trap. By 1992, they had come within a tantalizing factor of 3 of observing BEC but were stopped short due to technical problems.

Meanwhile, a parallel effort in observing BEC using alkali atoms was getting underway. The main impetus for this was to see if the tremendous developments that occurred in the late 1980s in using lasers to cool atomic clouds could be used to achieve BEC. Alkali atoms could be maintained in a gaseous state if the density was low, typically less than 10^{14} atoms/cm^3. But this meant that BEC would occur only at temperatures below 1 μK. Laser-cooling techniques had indeed achieved temperatures in the range of a few μK, with a corresponding increase in phase-space density of about 15 orders of magnitude. However, there were limitations in the achievable temperature due to heating from the presence of scattered photons in the cloud. One advance to this problem came from the MIT group of Dave Pritchard. His then post-doc, Wolfgang Ketterle, proposed using a special magneto-optic trap in which the coldest atoms get shelved in a dark state where they do not interact with the laser anymore. Since these atoms do not see the light, they do not get heated out of the trap. This helped improve the density by another order of magnitude, but BEC was still a factor of million away.

Pritchard's group at MIT also demonstrated magnetic trapping of sodium at around the same time. Pritchard and his student, Kris Helmerson, proposed a new technique for evaporative cooling in such a trap: RF-induced evaporation. Instead of lowering the magnetic field to cause the hottest atoms to escape, as was done in the spin-polarized hydrogen experiments, they proposed using an RF field tuned to flip the spin of the hottest atoms. The magnetic trap is a potential well for atoms whose spin is anti-parallel to the magnetic field, but is a potential hill for atoms whose spin is parallel. Therefore, once the spin of the atom is flipped, it would find itself on the side of a potential hill and slide out. The beauty of this technique is that the RF frequency determines which atoms get flipped, while the trapping fields remain unchanged. Pritchard's group was however unable to demonstrate evaporative cooling in their magnetic trap because the density was too low.

Laser cooling and evaporative cooling each had their limitations because they required different regimes. Laser cooling works best at low densities, while evaporative cooling works at high densities when collisions enable rapid rethermalization. Therefore, in the early 1990s, a few groups started using a hybrid approach to achieve BEC, i.e. first cool atoms to the microkelvin range using laser cooling, and then load them into a magnetic trap for evaporative cooling. By the year 1994, two groups were leading the race to obtain BEC: the Colorado group of Cornell and Wieman, and the MIT group of Ketterle. Both groups had demonstrated RF-induced evaporative cooling in a magnetic trap, but found that there was a new limitation, namely a hole in the bottom of the trap from which atoms leaked out. The hole was actually the field zero at the center of the trap. When atoms crossed this point, there was no field to keep the atom's spin aligned, so it could flip its spin and go into the untrapped state. As the cloud got colder, atoms spent more time near the hole and were quickly lost from the trap.

Ketterle's solution to plug the hole was to use a tightly focused Ar-ion laser beam at the trap center. The optical force from the laser beam kept the atoms out of this region, and, since the laser frequency was very far from the resonance frequency of the atoms, it did not cause any absorption or heating. The technique proved to be an immediate success and gave Ketterle's team an increase of about 3 orders of magnitude in phase-space density. But again technical problems limited the final observation of BEC.

Cornell had a different solution to the leaky trap problem: the time-orbiting potential (TOP) trap. His idea can be understood in the following way. The magnetic trap has a field whose magnitude increases linearly from zero as you move away from the trap center in any direction. The hole in the trap is the field zero point. Now, if you add a constant external field to this configuration, the hole does not disappear, it just moves to a new location depending on the strength and direction of the external field. Atoms will eventually find this new hole and leak out of it. However, Cornell's idea was that if you move the location of the hole faster than the average time taken for atoms to find it, the atoms will be constantly chasing the hole and never find it! A smooth way to achieve this is to add a rotating field that moves the hole in a circle. The time-averaged potential is then a smooth potential well with a non-zero minimum.

Plugging the leaky trap proved to be the final hurdle in achieving BEC. In July 1995, Cornell and Wieman announced that they had observed BEC in a gas of ^{87}Rb atoms. The transition temperature was a chilling 170 nK, making it the coldest point in the universe! The researchers had imaged the cloud by first allowing it to expand and then illuminating it with a pulse of resonant light. The light absorbed by the cloud cast a shadow on a CCD camera. The "darkness" of the shadow gave an estimate of the number of atoms in any region. The striking feature of the work was that there were three clear and distinct signatures of BEC, so clear that any skeptic would be immediately convinced:

1. The appearance of the condensate was marked by a narrow, intense peak of atoms near the center, corresponding to the ground state of the trap.

2. As the temperature was lowered below the transition temperature, the density of atoms in the peak increased abruptly, indicating a phase transition.

3. The atoms in the peak had a nonthermal velocity distribution as predicted by quantum mechanics for the ground state of the trap, thus indicating that all these atoms were in the same quantum state.

Soon after this, Ketterle's group observed BEC in a cloud of ^{23}Na atoms. As against the few thousand condensate atoms in the Colorado experiment, they had more than a million atoms in the condensate. This enabled them to do many quantitative experiments on the fundamental properties of the condensate. For example, they were able to show that when two condensates were combined, they formed an interference pattern, indicating that the atoms were all phase coherent. They were also able to extract a few atoms from the condensate at a time to form a primitive version of a pulsed atom laser: a beam of atoms that are in the same quantum state. They could excite collective modes in the condensate and watch the atoms slosh back and forth. These results matched the theoretical predictions very well.

BEC in atomic gases has since been achieved in several laboratories around the world. Apart from Rb and Na, it has been observed in all the other alkali atoms. The H group at MIT achieved it in 1998. Metastable He has also been cooled to the BEC limit. A Rb BEC has also been obtained by evaporative cooling in a crossed dipole trap, thus eliminating the need for strong magnetic fields and allowing atoms to be condensed independent of their spin state. Such an all-optical trap has been used to condense the two-electron atom Yb.

The variety of systems and techniques to get BEC promises many applications for condensates. The primary application, of course, is as a fertile testing ground for our understanding of many-body physics, bringing together the fields of atomic physics and condensed-matter physics. In precision measurements, the availability of a giant coherent atom should give enormous increase in sensitivity. BECs could also impact the field of nanotechnology since the ability to manipulate atoms greatly increases with their coherence.

Chapter 10

The 2005 Nobel Prize in Physics—Laser physics

THE 2005 Nobel Prize in Physics was awarded in the area of laser physics. One half of the prize (theory part) has been given to Roy Glauber of Harvard University, Cambridge (USA) *"for his contribution to the quantum theory of optical coherence"*, which became important soon after the invention of the laser. The second half of the prize (experimental part) was jointly awarded to two physicists, John Hall of the National Institute of Standards and Technology in Boulder (USA); and Theodor Hänsch of the Max Planck Institute for Quantum Optics, Garching (Germany). They have been cited *"for their contributions to the development of laser-based precision spectroscopy, including the optical frequency comb technique"*. In this chapter, we will discuss the contributions of Hall and Hänsch, which led to their being awarded the experimental part of the Prize.

A. Laser spectroscopy

Lasers have impacted our lives in countless number of ways. Today they are found everywhere—in computer hard disk drives, in CD players, in grocery store scanners, and in the surgeon's kit. In research laboratories, almost everyone uses lasers for one reason or another. However, arguably the greatest impact of lasers in physics has been in high-resolution spectroscopy of atoms and molecules. To see this, consider how spectroscopy was done before the advent of lasers. You would use a high-energy light source to excite all the transitions in the system, and then study the resulting emission "spectrum" as the atoms relaxed back to their ground state. This is like studying the modes of vibration of a box by hitting it with a sledgehammer and then separating the resulting sound into its different frequency components. A more gentle way of doing this would be to excite the system with a tuning fork of a given frequency. Then by changing the frequency of the tuning fork, one could build up the spectrum of the system. This is how you do laser spectroscopy with a (tunable) laser; you study the absorption of light by the atoms as you tune the laser frequency. When you come close to an atomic resonance, you build up an absorption curve with a characteristic width called the natural width.

In order to be able to do such high-resolution laser spectroscopy, two things have to be satisfied. First, the atomic resonance should not be artificially broadened. This can happen, for example, due to the Doppler effect in hot vapor, where the thermal velocity causes a frequency shift and broadens the line. Even with atoms at room temperature, the Doppler width can be 100 times the natural width, and can prevent closely spaced levels from being resolved. The second requirement for high-resolution spectroscopy is that the tunable laser should have a narrow "linewidth". The linewidth of the laser, or its frequency uncertainty, is like the width of the pen used to draw a curve on a sheet of paper. Obviously, you cannot draw a very fine curve if you have a broad pen.

It is in the above context that the Nobel citation mentions the work of the two laureates in laser-based precision spectroscopy. Their names are quite well known to anyone working in laser spectroscopy. In the early 1970s, Hänsch, then working at Stanford University with Arthur Schawlow (Nobel Prize for laser spectroscopy in 1981), pioneered the use of Doppler-free techniques such as saturation spectroscopy, particularly for spectroscopy in hydrogen. Around the same time, Hall developed many techniques to stabilize the frequency of lasers and reduce their linewidth. Today, two of the most popular techniques for laser stabilization are called the Hänsch-Couillaud technique and the Pound-Drever-Hall technique, in honor of these scientists.

In 1976, Hall and coworkers used high-resolution laser spectroscopy in methane to observe for the first time the recoil-induced splitting of a line. In other words, when the molecule absorbs a photon of wavelength λ, the photon momentum h/λ imparts a recoil to the molecule. This recoil velocity results in a frequency shift due to the Doppler effect. But this is a small effect, about 2 kHz in a frequency of 10^{14} Hz, and requires an extremely high resolving power. In the same year, Hänsch and Schawlow proposed that the momentum of laser photons could be used to cool atoms to very low temperatures—a technique that is now called "laser cooling" and which was discussed extensively in the last two chapters.

Many advances in physics have been brought about by high-resolution spectroscopy of atoms. Indeed, one might argue that the most obvious manifestation of quantization (or discreteness) at the atomic scale is the fact that atomic spectra show sharp spectral lines. The well-known Fraunhofer lines were first observed in the solar spectrum as dark lines using a spectrometer that was "high-resolution" for its time. In the early part of the twentieth century, Niels Bohr was able to explain such discrete lines by postulating that an electron in an atom was allowed only certain values of angular momentum. This led to the development of quantum mechanics as a theory in the atomic domain. Further measurements of atomic spectra at higher resolution revealed that many lines were actually doublets. A common example is the yellow light emitted by the ubiquitous sodium vapor lamp—it actually consists of two lines, called D_1 and D_2, which can be resolved and measured in a high-school laboratory today. The origin of this splitting is the interaction between two types of electronic angular momenta—orbital and spin. In 1928, Dirac wrote down his famous equation to describe the electron, which incorporated its spin angular momentum in a natural way. However, even the very successful Dirac theory predicted that the 2S and 2P states of hydrogen have the same energy. A precise measurement of these levels by Lamb showed that their energies are slightly different, which is now called the Lamb shift. The discovery of the Lamb shift led to the birth of quantum electrodynamics (QED), for which the Nobel Prize was awarded to Feynman, Schwinger, and Tomonaga in 1965.

B. The frequency comb

We thus see that improvement in precision almost always leads to new discoveries in physics. In recent times, one atomic transition that has inspired many advances in high-resolution spectroscopy and optical frequency measurements is the 1S → 2S resonance in hydrogen, with a natural width of only 1 Hz. Measurement of the frequency of this transition is important as a test of QED and for the measurement of fundamental constants. However, the frequency of this transition is 2.5×10^{15} Hz. Since the SI unit of time is defined in terms of the cesium radio-frequency transition at 9.2×10^9 Hz, measuring the optical frequency with reference to the atomic clock requires spanning 6 orders of magnitude! You can think of this as having two shafts whose rotation speeds differ by a factor of 1 million, and you need to measure the ratio of their speeds accurately. If we use a belt arrangement to couple the two shafts, then there is a possibility of errors in the ratio measurement due to phase slip. Instead, one would like to couple them through a gearbox mechanism with the correct teeth ratio so that there is no possibility of slip (see Fig. 10.1). This is precisely what is achieved by the frequency comb.

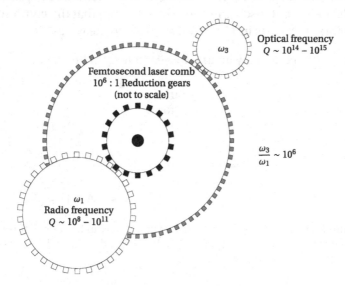

Figure 10.1: Schematic of a gearbox mechanism to couple radio frequencies to optical frequencies. The span of 6 orders of magnitude is what is achieved by the frequency comb.

The idea behind the comb technique, as seen in Fig. 10.2, is that periodicity in time implies periodicity in frequency. Thus, if you take a pulsed laser that produces a series of optical pulses at a fixed repetition rate, then the frequency spectrum of the laser will consist of a set of uniformly spaced peaks on either side of a central peak. The central peak is at the optical frequency of each laser pulse (called the carrier frequency), and the peaks on either side are spaced by the inverse of the repetition rate (called sidebands). You can produce such a spectrum by putting the laser through a nonlinear medium such as a nonlinear fiber. The larger the nonlinearity, the more the number of sidebands. Around 1999, there was a major development in making nonlinear fibers—fibers with honeycomb microstructure were developed which had such extreme nonlinearity that the sidebands spanned almost an octave. If you were to send a pulsed laser (operating near 800 nm) through such a fiber, you would get a near continuum of sidebands spanning the entire visible spectrum. The series of uniformly spaced peaks stretching out over a large frequency range looks like the teeth of a comb, hence the name *optical frequency comb*. The beautiful part of the technique is that the comb spacing is determined solely by the repetition rate, thus by referencing the repetition rate to a cesium atomic clock, the comb spacing can be determined as precisely as possible. In 1999, Hänsch and coworkers showed that the comb spacing was uniform to 3 parts in 10^{17}, even far out into the wings.

Periodicity in time $\Longleftrightarrow$ Periodicity in frequency

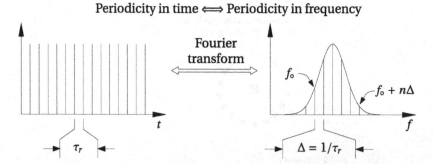

Figure 10.2: Periodicity in time implies periodicity in frequency through the Fourier transform. The spacing Δ in frequency domain is the inverse of the repetition period τ_r in time domain.

Thus the procedure to produce a frequency comb is now quite straightforward. One starts with a mode-locked, pulsed Ti:sapphire laser and sends its output through 20–30 cm of nonlinear fiber. The pulse repetition rate is referenced to an atomic clock, and determines the comb spacing. The carrier frequency is controlled independently, and determines the comb position.

But how does one measure an optical frequency using this comb? This can be done in two ways. One way is to use a reference transition whose frequency f_o is previously known. We now adjust the comb spacing Δ so that the reference frequency f_o lies on one peak, and the unknown frequency f lies on another peak that is n comb lines away [*]

$$f = f_o + n\Delta$$

Thus by measuring n, the number of comb lines in between, and using our knowledge of f_o and Δ, we can determine f. This was the method used by Hänsch in 1999 to determine the frequency of the D_1 line in cesium (at 895 nm). The measurement of this frequency can be related to the fine-structure constant α, which is one of the most important constants in physics because it sets the scale for electromagnetic interactions and is a fundamental parameter in QED calculations.

However, the above method requires that we already know some optical frequency f_o. If we want to determine the absolute value of f solely in terms of the atomic clock, the scheme is slightly more complicated. In effect, we take two multiples (or harmonics) of the laser frequency, and use the uniform comb lines as a precise ruler to span this frequency difference. Let us say we align one peak to $3.5f$, and another peak that is n comb lines away to $4f$, then we have determined

$$4f - 3.5f = n\Delta$$

which yields that

$$f = 2n\Delta$$

so that we have f in terms of the comb spacing. In 2000, Hänsch and coworkers used this method to determine the frequency of the hydrogen $1S \rightarrow 2S$ resonance with an unprecedented accuracy of 13 digits. This was the first time that a frequency comb was used to link a radio frequency to an optical frequency.

---

[*]It is not necessary that the comb peak aligns perfectly with the laser frequency. A small difference between the two can be measured easily since the beat signal will be at a sufficiently low frequency.

Currently, one of the most important questions in physics is whether fundamental constants of nature are really constant, or are they changing with time. For example, is the fine-structure constant α constant throughout the life of the universe or is it different in different epochs? Now, if you want to measure a very small rate of change $\dot{\alpha}$ (= $d\alpha/dt$), then you can do it in two ways. You can take a large dt so that the integrated change in α is very large. This is what is done in astronomy, where looking at the light from a distant star is like looking back millions of years in time. You can then compare atomic spectra from distant stars to spectra taken in the laboratory today. Alternately, if you want to do a laboratory experiment to determine $\dot{\alpha}$, then you have no choice but to use a small dt. Therefore, you have to improve the accuracy of measuring α so that even small changes become measurable. This is what has been done by Hänsch and his group. By measuring the hydrogen 1S $\rightarrow$ 2S resonance over a few years, they have been able to put a limit on the variation of α. Similar limits have been put by other groups using frequency-comb measurements of other optical transitions. The current limit on $\dot{\alpha}/\alpha$ from both astronomy and atomic physics measurements is about 10^{-15} per year.

In the last few years, several optical transitions have been measured using frequency combs. The primary motivation is to find a suitable candidate for an optical clock to replace the microwave transition used in the current definition. An optical clock will "tick" a million times faster, and will be inherently more accurate. However, since the cesium atomic clock has an accuracy of 10^{-15}, one has to measure the candidate optical transition to this accuracy to make sure it is consistent with the current definition. The race is on to find the best candidate among several alternatives such as laser-cooled single ions in a trap, ultracold neutral atoms in an optical lattice, or molecules. As seen from Fig. 5.1 of the "Standards" chapter, the accuracy of clocks has increased by several orders of magnitude in the last millennium, and is projected to increase further in the near future. The applications for more precise clocks of the future range from telecommunications and satellite navigation to fundamental physics issues such as measurement of pulsar periods, tests of general relativity, and variation of physical constants.

Chapter 11

The 2009 Nobel Prize in Physics—Achievements in optics that have changed modern life

THE winners of the 2009 Nobel prize in physics have been recognized for two achievements in optics that have revolutionized modern life in the past few decades. The first is the invention of the optical fiber, which makes long-distance communication literally lightning fast. Our modern life is fundamentally dependent on optical communication, from the Internet to the ubiquitous mobile phone. The second invention is the CCD (short form of *charge-coupled device*) camera, the electronic eye that has changed optical imaging and put the traditional photographic film out of use. Today, even the cheapest cell phone has a high-resolution (mega-pixel) CCD camera that allows you to take a picture and email it to a friend halfway across the world. A child growing up in this interconnected world would find it hard to believe that these things were in the realm of science fiction just a short generation ago.

A. The optical fiber

All of us learnt in school that light travels in straight lines—the so-called rectilinear propagation of light. Imagine if you could make a pipe for light and make it flow along this pipe like water in a water pipe, whatever be the curves and bends along the path. Well that is exactly what an *optical fiber* does. It is a thin wire of glass that confines the light and transports it along its length. And this length is not just a short distance of a few meters, but extends over thousands of kilometers.

So how is the fiber able to do this? It uses the idea of *total internal reflection*, the well-known phenomenon that light, incident on an interface where the refractive index changes from high to low, will undergo reflection if the angle of incidence is larger than a critical angle. From Snell's law of refraction

$$\frac{\sin \theta_1}{\sin \theta_2} = \frac{n_2}{n_1}$$

where θ_1 is the angle of incidence in medium 1, θ_2 is the angle of refraction in medium 2, and n_1 and n_2 are the refractive indices of the two media. Both angles are measured with respect to the normal at the boundary. If $n_1 > n_2$, there is a critical angle θ_c above which there is no real value of θ_2 that can satisfy the above equation. The light then undergoes total internal reflection, the phenomenon that gives a cut diamond its sparkle. It is called *total* because at smaller angles of incidence these is partial reflection and partial refraction, and the reflection becomes total beyond θ_c. For the glass-to-air interface used in optical fibers, the refractive index changes from 1.5 to 1 and hence the critical angle is 41.8°.

Guiding of light using a region of high refractive index is an old idea. It was first demonstrated in a stream of water by Daniel Colladon and Jacques Babinet in Paris in the early 1840s. The modern optical fiber, made entirely of glass and consisting of a high-index core surrounded by a low-index cladding, was invented in the 1950s. Experiments conducted by the American physicist, Narinder Singh Kapany, played an important role in this development. Indeed, Kapany is rightly acknowledged as the father of the optical fiber and coined the term *fiber-optic*. But these early fibers suffered from severe attenuation of the light signal as it propagated through the fiber, making it impractical for long-distance communication.

This is where the breakthrough contribution of Kao comes in. In 1966, when he was working for the British company Standard Telephones and Cables, he published a seminal paper along with his colleague G. A. Hockam. In the paper, titled "*Dielectric-fibre surface waveguides for optical frequencies*", they proposed that the attenuation in fibers was caused by impurities which could be eliminated, rather than fundamental physical effects such as scattering. They correctly pointed out that fibers with low loss could be manufactured by using high-purity glass. The benchmark at the time was an attenuation level of 20 decibel per kilometer (dB/km).* Indeed, just 4 years later, researchers at the American company Corning demonstrated a fiber with an attenuation of only 17 dB/km by doping silica glass with titanium. A few years later they produced a fiber with only 4 dB/km attenuation using germanium dioxide as the core dopant. Such low attenuations made optical fiber telecommunications a practical reality.

Today, optical fibers are used everywhere from scientific labs to the cable that brings TV and Internet to your home. In many labs, *single-mode* fibers are used to transport laser light from one point to another. Apart from allowing light to be transported without the use of steering mirrors, the main advantage of using a single-mode fiber is that the mode coming out of the fiber is pure Gaussian. The light coming out of a laser is Gaussian to start with, but gets distorted due to optical elements and even dust particles along the way; the use of a single mode fiber is a great way to clean it up.

One of the most important uses of optical fibers is in *endoscopy*, the minimally invasive technique used in medicine to image internal parts of the human body. The endoscope usually consists of a flexible fiber bundle and a small camera lens at the end. The fibers both deliver light to illuminate the object and bring the scattered light back to form the image. Endoscopy can help a surgeon visualize hard-to-reach areas of, for example, the gastro-intestinal tract, and even deliver the medication with the same endoscope if needed.

No discussion of optical fibers would be complete without a mention of *solitons*, solitary waves that propagate without changing shape. In 1973, Akira Hasegawa, working at Bell Labs of the American telephone company

---

*The decibel unit, defined as $\{-10\log(P_2/P_1)\}$, is a logarithmic scale. Thus, a 20 dB attenuation would correspond to a reduction in power by a factor of 10^{+2}, or that 1% of the input power survives.

AT&T, was the first to suggest that solitons could exist in nonlinear optical fibers, due to a balance between self-phase modulation and anomalous dispersion. In 1988, Linn Mollenauer and his team, again at AT&T Bell Labs, transmitted soliton pulses over 4000 kilometers using the Raman effect. AT&T immediately decided to invest billions of dollars laying trans-Atlantic fiber-optic cables instead of the conventional copper cables, because they were convinced that the future of communication was optical. Time has proved this hunch right. While solitons are not in common use now, their use in the near future may increase the bandwidth of existing optical cables many fold.

Fiber-optic cables are at the heart of today's communication-intensive world. Such cables are capable of transmitting trillions of bits of information every second, bits of information carrying voice, data, and still and video images. Since the time of Edison, we know how to convert audio into electrical signals; but converting an image to electronic form, that is a different matter altogether. Which brings us to the second part of the Nobel Prize—the CCD sensor.

B. The CCD camera

A conventional camera consists of an optical system that forms an image on the image (or focal) plane, and a film at the image plane where some photosensitive material changes its properties depending on the amount of light incident at any point. Upon developing, the film is transformed into a photograph that represents the object which was imaged. The CCD camera is an electronic version of the same idea, with the photographic film at the image plane being replaced by a CCD array.

Since we want an electronic image, we have to first convert the photons into electrons. This is done by making use of the photoelectric effect, first explained by Albert Einstein in 1905. Einstein correctly pointed out that the number of electrons produced in this process is proportional to the intensity of the light. Therefore, if we could design a sensor that could collect and read out the number of electrons at a given image point, we would have an electronic camera. Each image point is called a *pixel*, and the larger the number of pixels in a given area, the better the resolution. So the challenge was to read a large number of pixels in a short time.

This is what the CCD sensor invented by Boyle and Smith in 1969 achieved. Think of each pixel as a bucket that can hold a certain number of electrons. The number of electrons or charge accumulated in each bucket is proportional to the intensity of light incident on the pixel. The pattern is read out as a shift register:

1. the last bucket is read out first,

2. each bucket then transfers its charge to the next neighboring bucket and the last one is read out again,

3. the process is repeated until all the buckets are empty.

The key to the design is the ability of the charge to be transferred from one bucket to the next, hence the name *charge-coupled device*. In practice, each bucket is a silicon capacitor. Since the charge Q in a capacitor C is linearly related to the voltage V,

$$Q = CV$$

the charge can be transferred between neighboring capacitors by suitably changing the two voltages. Similarly, the charge in the last capacitor is transferred to a charge amplifier and read out as a voltage. Thus, the entire contents of the CCD array are converted to a sequence of voltages. If the array is two-dimensional, we have our electronic image.

Boyle and Smith were then working at AT&T Bell labs, the same place mentioned earlier in connection with the soliton work. Until the 90's, Bell Labs provided an unparalleled intellectual environment for doing basic research. It is a shame that such industrial research labs are becoming extinct now. Boyle and Smith reminisce fondly about the freedom they had in their work and the lack of any "directives" from upper management. This was partly because the head of the lab was not a money-minded bureaucrat but a scientist like them. There was an open management style with no hierarchy, and senior scientists would stop by for casual chats. It is no wonder that Bell Labs has produced about 10 Nobel laureates in physics, starting with C. J. Davisson in 1937 for the famous Davisson-Germer experiment which confirmed the de Broglie hypothesis on the wave nature of matter.

The CCD camera has made photography accessible to everyone. Each pixel in a CCD array is only about 10 μm in size, therefore a chip of few square cm can hold several million pixels. Small cameras and cell phones boast of a few mega pixels, which gives a resolution unheard of a decade ago. The camera takes such high-resolution images that you can zoom into any part of the image without significant loss in resolution. And you can take poster-size prints of any photo that you like.

But the use of the CCD camera is not limited to the lay public. It plays an equally important role in science and technology. In fact, at the front end of most endoscopes mentioned earlier is a CCD camera that allows the surgeon to see the images in real time. In atomic physics labs, CCD cameras are used to image trapped particles. They also find widespread use in astronomical telescopes due to their high quantum efficiency and linearity (i.e. signal proportional to the light intensity). The photoactive part of the CCD array can easily be made sensitive to different regions of the electromagnetic spectrum, ranging from X-rays to UV to visible to infrared. Each one gives us a different eye into the universe.

Chapter 12

The 2012 Nobel Prize in Physics—Manipulation of single particles

Q UANTUM mechanics describes the weird world of microscopic particles, which is completely different from the macro world that we are used to in our daily lives. Perhaps the most defining characteristic of that world is that it is *discrete*—hence the name *quantum*—as opposed to the continuous nature of physics at the macrolevel. The discreteness of the microscopic world was first explained by Planck in 1900 with his derivation of the blackbody spectrum. Since then, we have understood that discreteness is an inherent feature of both matter (atoms, or a collection of them) and its interactions (light in the form of photons, for example).

The 2012 Nobel laureates—David Wineland of the National Institute of Standards and Technology, Boulder (USA); and Serge Haroche of College de France (France)—have been honored for experiments that demonstrate the "quantum" behavior of nature at the microlevel. In this chapter, we will discuss some of their experiments that demonstrate this discreteness.

A. Trapped ions

From the time of the Bohr, we know that atoms have discrete energy levels, and can make a transition from one level to another—a so-called *quantum jump*. If the first level is an excited state, and the second level is the ground state, the transition is through the process of spontaneous emission with a rate determined by the lifetime of the upper state. What Wineland and his team showed in 1986 was the observation of a quantum jump of a single ion held in a Paul trap. The Paul trap uses radio-frequency fields to confine a charged particle—a single Hg^+ ion in this case. The ground state of the ion is an S state, and the first resonance line is the S $\rightarrow$ P transition, which is driven by a laser. There is also a metastable D state that lies in between the two levels. It is metastable (or long-lived) because the transition to the ground state is forbidden by selection rules.

The experimental demonstration of quantum jumps now proceeds as follows. The ion is irradiated with laser light driving the first resonance line. As it absorbs and emits the light, the ion shines with resonance fluorescence, which can be imaged with a CCD camera or some photodetector. The ion in this condition is called "bright". The signal is nearly continuous because the lifetime of the excited P state is only about 2 ns, much faster than the response time of any photodetector. Once in a while, the ion goes into the metastable D state (with a long lifetime of order 0.1 s), and becomes "dark". Dark because the ion in the metastable state cannot absorb the first-resonance radiation, and the fluorescence signal goes off. The signal comes on again when the ion decays back to the ground state. Therefore, the fluorescence signal as a function of time shows on and off periods, with *sudden transitions* between the two—a direct demonstration of quantum jumps. The statistical average of all the off periods over a long observation time gives the lifetime of the D state.

The ion used in the above study was *laser cooled*. The idea of laser cooling has been described in detail in the chapter on the 1997 Nobel Prize. In a trap, this process is called *sideband cooling*, as discussed in the chapter on Robert Dicke. Wineland's group was the first to demonstrate laser cooling of ions held in a trap. They also demonstrated a precise atomic clock based on a single laser-cooled trapped ion.

The experiments of Wineland show elegantly that a single trapped ion is useful for many experiments—both fundamental and applied. The latest application is in the field of quantum computing. The bits in a quantum computer—called *qubits*—can take on any superposition of 0 and 1, while that of a classical computer can only be one of the two. This feature, along with other quantum mechanical aspects like entanglement, can be exploited to compute some things much faster than can be done classically. The trapped laser-cooled ion is an almost ideal qubit because it is a single particle that is free from perturbations, and its energy levels can be addressed very precisely using lasers. The 0 and 1 state of the qubit are the ground and a metastable state, respectively. Ca^+ ions in a linear Paul trap are now the species of choice for such applications because the laser cooling transition is accessible with a low-cost diode laser. Therefore, several groups worldwide are doing experiments on Ca^+ ions for quantum computing applications.

B. Single photons

So far we have seen how to use photons to manipulate single particles of matter. Haroche's experiments are complementary in the sense that he and his team use atoms to manipulate single photons, thereby demonstrating the quantum nature of light. The quantized EM field is like quantizing a harmonic oscillator, except that instead of the position and velocity oscillating in a harmonic oscillator, it is the electric and magnetic fields that oscillate in a light wave. The detailed theory for getting quantized photons from the classical fields is called *quantum electrodynamics* (QED).

For understanding the experiments of Haroche, understanding two features of their experimental scheme are important. One is that they use *circular Rydberg* states. These are states that have a large value of the principal quantum number n (about 50), and the largest possible value of the orbital quantum number ℓ

$$\ell = n - 1$$

An atom in this state is circular in the sense that the valence electron is represented by a wave packet in a circular orbit. The nucleus and the inner electrons are far enough away that the potential seen by the outer electron is similar to that in the hydrogen atom, and the simple Bohr model can be used to explain most of its features. Some of these features are that the atom has a very large magnetic moment, and a long radiative lifetime for transitions.

The second important aspect of their experiments is the use of *cavity quantum electrodynamics* (cavity QED). The transition that they use is at 51.1 GHz from $n = 50$ to $n = 51$, accessed inside a superconducting microwave cavity—two niobium spherical mirrors in a Fabry-Perot configuration cooled to a temperature of about 1 K. At this temperature, the number of thermal photons is about 0.1. It is an extremely high Q cavity, so that a photon bounces back and forth between the mirrors for 1 ms before decaying. More importantly, the interaction of the Rydberg atom with the cavity is so strong that it overwhelms all dissipative coupling to the environment. And the radiative lifetime of the transition is 30 ms, much longer than other time scales.

The non-destructive measurement of a single photon proceeds as follows. Rydberg atoms are sent through the microwave cavity. The cycle of photon absorption and emission during the interaction causes a shift in the phase of the atomic wavefunction, the so-called Rabi oscillation. The cavity is first *prepared* by sending an atom in the upper state. The interaction time is adjusted for a phase shift of $\pi/2$, corresponding to a quarter Rabi oscillation: a $\pi/2$ pulse in the terminology of NMR. It leaves the cavity in a superposition state with 0 or 1 photon — 0 if the atom is in the upper state, and 1 if the atom is in the lower state. A second "meter" atom performs a *non-demolition* measurement of the state of the cavity. The interaction time is now adjusted for a phase shift of 2π, i.e. a full Rabi oscillation. The measurement is non-destructive because the state of the cavity is the same before and after the measurement. The phase of the meter atom is measured using fringes in an atom interferometer, with a set of Ramsey SOF pulses. The fringe pattern shifts by π depending on whether the cavity has 1 photon or 0. The interaction with 0 photons is with the vacuum modes of the quantized electromagnetic field, as discussed in the "Oscillations" chapter. To reiterate, the mean value of the E and B fields is zero, but their mean squared value is not. The interaction with this vacuum causes what is called a *vacuum Rabi oscillation.*

We can now see that this same set up can be used to perform an analogue of the quantum jump, but this time for photons. The cavity is used to store one microwave photon for a long time. The meter atom shows a Ramsey fringe pattern corresponding to the presence of one photon. The fringe pattern shifts *suddenly* when the photon dies, a quantum jump in the life of the photon. This is what Haroche and his group showed in 2007, nicely complementing the quantum jump experiments of Wineland with atoms.

Index

Printed in the United States
By Bookmasters